Mathematical Olympiads 1998–1999

Problems and Solutions From Around the World

Edited by
Titu Andreescu
and
Zuming Feng

Published and distributed by
The Mathematical Association of America

MAA PROBLEM BOOKS SERIES

Problem Books is a series of the Mathematical Association of America consisting of collections of problems and solutions from annual mathematical competitions; compilations of problems (including unsolved problems) specific to particular branches of mathematics; books on the art and practice of problem solving, etc.

The Inquisitive Problem Solver, Paul Vaderlind, Richard K. Guy, and Loren L. Larson

Mathematical Olympiads 1998–1999: Problems and Solutions From Around the World, edited by Titu Andreescu and Zuming Feng

Mathematical Olympiads 1999–2000: Problems and Solutions From Around the World, edited by Titu Andreescu and Zuming Feng

The William Lowell Putnam Mathematical Competition 1985–2000: Problems, Solutions, and Commentary, Kiran S. Kedlaya, Bjorn Poonen, Ravi Vakil

USA and International Mathematical Olympiads 2000, edited by Titu Andreescu and Zuming Feng

USA and International Mathematical Olympiads 2001, edited by Titu Andreescu and Zuming Feng

MAA Service Center
P. O. Box 91112
Washington, DC 20090-1112
1-800-331-1622 fax: 1-301-206-9789

Mathematical Olympiads 1998–1999

Problems and Solutions From Around the World

Library of Congress Catalog Card Number 00-105112
ISBN 0-88385-803-7
Printed in the United States of America
Current Printing (last digit):
10 9 8 7 6 5 4 3 2

Contents

Preface

This book is a continuation of *Mathematical Contests 1997–1998: Olympiad Problems and Solutions from around the World*, published by the American Mathematics Competitions. It contains solutions to the problems from 25 national and regional contests featured in the earlier book, together with selected problems (without solutions) from national and regional contests given during 1999. In many cases multiple solutions are provided in order to encourage students to compare different problem-solving strategies.

This collection is intended as practice for the serious student who wishes to improve his or her performance on the USAMO. Some of the problems are comparable to the USAMO in that they came from national contests. Others are harder, as some countries first have a national Olympiad, and later one or more exams to select a team for the IMO. And some problems come from regional international contests ("mini-IMOs").

Different nations have different mathematical cultures, so you will find some of these problems extremely hard and some rather easy. We have tried to present a wide variety of problems, especially from those countries that have often done well at the IMO.

Each contest has its own time limit. We have not furnished this information, because we have not always included complete exams. As a rule of thumb, most contests allow time ranging between one-half to one full hour per problem.

The problems themselves should provide much enjoyment for all those fascinated by solving challenging mathematics questions.

Acknowledgments

Thanks to Kiran Kedlaya for the help he provided in ways too numerous to mention, and to the following students of the 1999 Mathematical Olympiad Summer Program who helped in preparing and proofreading solutions: George Lee, Reid Barton, Gabriel Carroll, Po-Shen Loh, Po-Ru Loh, Yuran Lu, Dani Kane, Zhihao Liu. Without their efforts this work would not have been possible. And finally thanks to Richard Parris, Andreas Orphanides, and Juan Felix for translating problems, and to Eugene Mukhin for providing hints.

Titu Andreescu Zuming Feng

Abbreviations and Notations

Abbreviations

AIME	American Invitational Mathematics Examination
APMC	Austrian-Polish Mathematics Competition
APMO	Asian Pacific Mathematics Olympiad
IMO	International Mathematical Olympiad
USAMO	United States of America Mathematical Olympiad
MOSP	Mathematical Olympiad Summer Program
LHS	Left-hand side
RHS	Right-hand side
AM-GM	Arithmetic-Geometric Mean Inequality
RMS-AM	Root Mean Square-Arithmetic Mean Inequality
WLOG	Without loss of generality

Notations for Numerical Sets and Fields

$\mathbb{Z}$	the set of integers
$\mathbb{Z}^+$	the set of positive integers
$\mathbb{Z}_n$	the set of integers modulo n
$\mathbb{Q}$	the set of rational numbers
$\mathbb{Q}^+$	the set of positive rational numbers
$\mathbb{Q}^n$	the set of n-tuples of rational numbers
$\mathbb{R}$	the set of real numbers
$\mathbb{R}^+$	the set of positive real numbers
$\mathbb{R}^n$	the set of n-tuples of real numbers
$\mathbb{C}$	the set of complex numbers
$[x^n](p(x))$	the coefficient of the term x^n in the polynomial $p(x)$
$\deg p$	the degree of the polynomial $p(x)$
$\deg(K/k)$	the degree of the field extension K/k

$\mathbb{F}[x]$	the ring of all polynomials with coefficients in the field $\mathbb{F}$.
$\mathbb{Q}(r)$	the field of complex numbers $g(r)/h(r)$ where $g(x), h(x) \in \mathbb{Q}[x]$ and $h(r) \neq 0$

Notations for Sets, Logic, and Geometry

$\Longleftrightarrow$	if and only if
$\Longrightarrow$	implies
$A \subset B$	A is a proper subset of B
$A \subseteq B$	A is a subset of B
$A \setminus B$	A without B
$A \cap B$	the intersection of sets A and B
$A \cup B$	the union of sets A and B
$a \in A$	the element a belongs to the set A
AB	segment AB; also the length of the segment AB
$\overrightarrow{AB}$	vector AB
$[\mathcal{F}]$	area of figure $\mathcal{F}$

1

1998 National Contests: Problems and Solutions

1.1 Bulgaria

Problem 1

Find the least natural number n ($n \geq 3$) with the following property: for any coloring in 2 colors of n distinct collinear points A_1, A_2, $\ldots$, A_n there exist three points A_i, A_j, A_{2j-i}, $1 \leq i < 2j - i \leq n$, which are colored in the same color.

Solution. The answer is 9. We call three points as stated in the question a *triple.* Let us denote the color of A_i by c_i, and let $c_i = r$ if A_i is red and $c_i = b$ if A_i is blue. We first exhibit a coloring of 8 points that contains no triple:

$$(c_1, c_2, c_3, c_4, c_5, c_6, c_7, c_8) = (r, b, r, b, b, r, b, r).$$

We claim that if $n \geq 9$, then there is at least one triple, i.e., there exists an arithmetic sequence i_1, i_2, i_3 such that $c_{i_1} = c_{i_2} = c_{i_3}$. Suppose, on the contrary, that there are no such triples in $A_1, \ldots, A_9$. At least 5 of them have the same color. For brevity, we abbreviate "the points numbered x, y, and z, where x, y, z form an arithmetic series, cannot all have the same color" with the symbol (x, y, z). WLOG, let $r = c_{i_1} = c_{i_2} = \cdots = c_{i_5}$, $i_1 < i_2 < \cdots < i_5$, be the first 5 red points. Since there is no triple, $i_2 - i_1 \neq i_3 - i_2$ and $i_3 - i_1 \geq 3$. Similarly, $i_5 - i_3 \geq 3$. Also, $i_3 - i_1 \neq i_5 - i_3$, so $i_5 - i_1 \geq 7$ and $(i_1, i_5) = (1, 8)$, $(1, 9)$, or $(2, 9)$. We consider the following three cases.

1. $(i_1, i_5) = (1, 8)$. Then $i_3 = 4$ or 5. By symmetry, WLOG, let $i_3 = 4$. Since $(1, 4, 7)$ and $(4, 6, 8)$, $i_4 = 5$. But then $(2, 5, 8)$ and $(1, 3, 5)$ will leave i_2 with no value to take, a contradiction.

2. $(i_1, i_5) = (1, 9)$. Then $i_3 = 4, 5$, or 6. Since $(1, 5, 9)$, $i_3 = 4$ or 6. By symmetry, WLOG, let $i_3 = 4$. Then $i_4 \neq 7$ and $i_2 = 2$ or 3. Since $(2, 4, 6)$ and $(3, 6, 9)$, $i_4 \neq 6$ and $i_4 = 8$. Then $c_5 = c_6 = c_7 = b$, also a contradiction.
3. $(i_1, i_5) = (2, 9)$. This is the same as (a). (To see this, replace every number x in (a) by $x + 1$.)

From above, we see that our assumption is false and there exists a triple in the first 9 points.

Problem 2

A convex quadrilateral $ABCD$ has $AD = CD$ and $\angle DAB = \angle ABC < 90^\circ$. The line through D and the midpoint of BC intersects line AB in point E. Prove that $\angle BEC = \angle DAC$. (Note: The problem is valid without the assumption $\angle ABC < 90^\circ$.)

Solution. Let M be the midpoint of BC, and let

$$\angle DAC = \angle DCA = x \text{ and } \angle CAB = y.$$

We have

$$\angle CBA = x + y,\ \angle ACB = 180^\circ - x - 2y,\ \angle DCB = 180^\circ - 2y.$$

Applying the law of sines to triangles BEM and ADE, we have

$$\frac{MB}{ME} = \frac{\sin \angle MEB}{\sin \angle EBM} = \frac{\sin \angle MEB}{\sin \angle EAD} = \frac{AD}{ED}.$$

Since $MB = MC$ and $AD = CD$, $MC/ME = CD/ED$. Applying the law of sines to triangles CEM and CDE, we have

$$\frac{\sin \angle CEM}{\sin \angle BCE} = \frac{MC}{ME} = \frac{CD}{ED} = \frac{\sin \angle CED}{\sin \angle DCE} = \frac{\sin \angle CEM}{\sin(180^\circ - \angle DCE)}.$$

So

$$\angle BCE = 180^\circ - \angle DCE = 180^\circ - \angle DCB - \angle BCE = 2y - \angle BCE,$$

which implies that $\angle BCE = y$. Hence

$$\angle CEB = \angle CBA - \angle BCE = x = \angle DAC,$$

as desired.

Problem 3

Let $\mathbb{R}^+$ be the set of positive real numbers. Prove that there does not exist a function $f : \mathbb{R}^+ \to \mathbb{R}^+$ such that

$$(f(x))^2 \geq f(x+y)(f(x)+y)$$

for every $x, y \in \mathbb{R}^+$.

First Solution. We proceed indirectly; suppose that f exists. Then,

$$f(x+y) \leq \frac{f(x)^2}{f(x)+y} < f(x).$$

So, f is a strictly decreasing function. (In particular, this means that f is one-to-one.)

But since $f(x)$ is always positive, f must have some fixed point; suppose that $f(a) = a$ for some a. Now we can plug a in for x and obtain: $f(a+y) \leq a^2/(a+y)$. This means that for some b, all $x \geq b$ will have $f(x) < 1$. Observe that if we plug $f(x)$ in for y in the given equation, then we have $f(x) \geq 2f(x+f(x))$. But now if we plug b in for x here, we find

$$1 > f(b) \geq 2f(b+f(b)) \geq 2f(b+1).$$

Hence $f(b+1) < 1/2$. Plugging in $b+1$ yields

$$1/2 > f(b+1) \geq 2f(b+1+f(b+1)) > 2f(b+1+1/2).$$

Hence $f(b+1+1/2) < 1/4$.

Proceeding in this manner, we find that

$$f\left(b + \sum_{i=0}^{n-1} 1/2^i\right) < 1/2^n.$$

Therefore, for each quantity $\epsilon \in (0,1)$, we can find a $\delta \in (0,2)$ such that $f(b+x) < \epsilon$ for all $x < 2-\delta$. This means that $f(x)$ does not exist for $x \geq b+2$, and so we are done.

Second Solution. This is another way of presenting the idea in the first solution.

Again, we proceed indirectly; suppose that f exists. As in the first solution, we can prove that f is a strictly decreasing function. For $(x,y) = (x, f(x))$, we have

$$f(x+f(x)) \leq \frac{f(x)^2}{f(x)+f(x)} = \frac{f(x)}{2}.$$

Let $x_0 \in \mathbb{R}^+$ and $f(x_0) = a$. For $(x, y) = (x_0, f(x_0))$, we have

$$\frac{a}{2} = \frac{f(x_0)}{2} \geq f(x_0 + f(x_0)) = f(x_0 + a).$$

Let $x_1 = x_0 + a$. Then $f(x_1) \leq a/2$. For $(x, y) = (x_1, f(x_1))$, we have

$$\frac{a}{4} \geq \frac{f(x_1)}{2} \geq f(x_1 + f(x_1)) \geq f\left(x_1 + \frac{a}{2}\right).$$

Let $x_2 = x_1 + a/2$. Then $f(x_2) \leq a/4$. For $(x, y) = (x_2, f(x_2))$, we have

$$\frac{a}{8} \geq \frac{f(x_2)}{2} \geq f(x_2 + f(x_2)) \geq f\left(x_2 + \frac{a}{4}\right).$$

Proceeding in this manner, we find that, as $n \longrightarrow \infty$,

$$f(x_0 + 3a) < f\left(x_0 + \sum_{i=0}^{n} \frac{a}{2^n}\right) < \frac{a}{2^{n+1}} \longrightarrow 0,$$

a contradiction. So our assumption is false and there is no such function.

Problem 4

Let $f(x) = x^3 - 3x + 1$. Find the number of distinct real roots of the equation $f(f(x)) = 0$.

First Solution. The answer is 7. We use calculus; the derivative of $f(x)$ is $3x^2 - 3$, and so the critical points are at $(-1, 3)$ and $(1, -1)$. But $f(-2) = -1, f(0) = 1$, $f(2) = 3$. So, the zeros of f lie in the intervals $(-2, -1), (0, 1)$, and $(1, 2)$. If we count the number of times f passes completely through each interval, that will be our answer. In fact, f passes through $(-2, -1)$ once (when $x < -2$), it passes through $(0, 1)$ three times (when $-2 < x < -1, 0 < x < 1$, and $1 < x < 2$), and it passes through $(1, 2)$ three times as well. Therefore there are a total of 7 roots for $f(f(x)) = 0$.

Second Solution. We begin with the calculus argument above to discover that $-2 \leq x \leq 2$. Now, let $x = 2\sin\alpha$. Then $x^3 - 3x = 2\sin 3\alpha$. We leave the details as an exercise.

Problem 5

The convex pentagon $ABCDE$ is inscribed in a circle of radius R. The radius of the incircle of triangle XYZ is denoted by r_{XYZ}. Prove that

1. $\cos\angle CAB + \cos\angle ABC + \cos\angle BCA = 1 + \dfrac{r_{ABC}}{R}$;

2. if $r_{ABC} = r_{AED}$ and $r_{ABD} = r_{AEC}$, then triangles ABC and AED are congruent.

Solution. **(a)** (For an alternate solution, please see APMC98 problem 9.) Let O and I be the circumcenter and incenter of triangle ABC; let $\angle CAB = \alpha, \angle ABC = \beta, \angle BCA = \gamma$. Let AI meet the circumcircle again at F. WLOG, let O lie inside $\angle AFC$. Then

$$OF = R, FI = FC = 2R\sin\frac{\alpha}{2},$$

and $\angle OFI = \beta + \alpha/2 - 90°$. Applying the law of cosines to triangle OFI, we have

$$\begin{aligned} OI^2 &= R^2 + 4R^2\sin^2\alpha/2 - 4R^2\sin(\alpha/2)\sin(\beta+\alpha/2) \\ &= R^2(4\sin^2\alpha/2 + 1 - 4\sin(\alpha/2)\sin(\beta+\alpha/2)) \\ &= R^2(2 - 2\cos\alpha + 1 + 2[\cos(\alpha+\beta) - \cos(\beta)]) \\ &= R^2(3 - 2\cos\alpha - 2\cos\beta - 2\cos\gamma). \end{aligned}$$

But $OI^2 = R^2 - 2Rr_{ABC}$ (Euler's formula), so

$$\cos\alpha + \cos\beta + \cos\gamma = 1 + \frac{r_{ABC}}{R}$$

and we are done.

(b) Let $2a, 2b, 2c, 2d, 2e$ be the measures of arcs AB, BC, CD, DE, EA, respectively. From (a), we have

$$\cos a - \cos(a+b) + \cos b = \cos d + \cos e - \cos(d+e),$$

$$\cos a + \cos(b+c) + \cos(d+e) = \cos e + \cos(c+d) + \cos(a+b).$$

Subtracting the two equations, we obtain:

$$\begin{aligned} &\cos b + \cos(c+d) = \cos d + \cos(b+c) \\ &\iff 2\cos\frac{b+c+d}{2}\cos\frac{b-c-d}{2} = 2\cos\frac{b+c+d}{2}\cos\frac{d-b-c}{2} \\ &\iff \cos\frac{b-c-d}{2} = \cos\frac{d-b-c}{2} \\ &\iff b = d. \end{aligned}$$

Plugging this result into the first equation, we obtain:

$$\begin{aligned} &\cos a - \cos(a+b) + \cos b = \cos d + \cos e - \cos(d+e) \\ &\iff \cos a - \cos(a+b) + \cos b = \cos b + \cos e - \cos(b+e) \end{aligned}$$

$$\Longleftrightarrow \cos a + \cos(b+e) = \cos e + \cos(a+b)$$

$$\Longleftrightarrow 2\cos\frac{a+b+e}{2}\cos\frac{a-b-e}{2} = 2\cos\frac{a+b+e}{2}\cos\frac{e-a-b}{2}$$

$$\Longleftrightarrow \cos\frac{a-b-e}{2} = \cos\frac{e-a-b}{2}$$

$$\Longleftrightarrow a = e.$$

Since $a = e$ and $b = d$, triangles ABC and AED are congruent.

Problem 6

Prove that the equation

$$x^2y^2 = z^2(z^2 - x^2 - y^2)$$

has no solutions in positive integers.

Solution. We begin by proving the following lemma.

Lemma 1. *The equation*

$$s^4 - t^4 = u^2 \tag{1}$$

has no solutions in positive integers.

Proof. We proceed indirectly; suppose that there exists a nonempty set S of integers such that if $a \in S$ then there exist natural numbers b and c such that $a^4 - b^4 = c^2$. By the Well Ordering Principle, there exists a minimum element of S; suppose that in our equation $a^4 - b^4 = c^2$, a is equal to the minimum. Thus

$$\gcd(a,b,c) = \gcd(a,b) = \gcd(b,c) = \gcd(c,a) = 1.$$

We consider the following cases.

(a) c is even. Then a and b are odd and $a^2+b^2 \equiv 2 \pmod 4$. Since $\gcd(a,b) = 1$, $4|c^2$, and

$$(a^2+b^2)(a^2-b^2) = c^2, \quad \gcd(a^2+b^2, a^2-b^2) = 2.$$

Let $x = \sqrt{(a^2+b^2)/2}$ and $y = \sqrt{(a^2-b^2)/2}$. Then x and y are both integers, and we have $x^4 - y^4 = (ab)^2$. Since $a > b$, $x < a$; therefore this violates our assumption that our choice of a yields a minimal solution.

(b) c is odd. Then $c^2 \equiv 1 \pmod 4$. Since $a^4, b^4 \equiv 0$ or $1 \pmod 4$, a is odd and b is even. We rewrite our equation: $(a^2)^2 = (b^2)^2 + c^2$. We use Pythagorean substitution. There exist positive integers r and s such that $\gcd(r,s)=1$,

$$a^2 = r^2 + s^2, \qquad b^2 = 2rs, \qquad c = r^2 - s^2.$$

We are only going to use the first two identities. (Both are symmetric with respect to r and s.) Since $2|b$, $4|b^2$ and exactly one of r and s is even. We denote by r' the even one and by s' the odd one. Then there exist positive integers x, y for which $r'/2 = x^2$ and $s' = y^2$. Thus $a^2 = 4x^4 + y^4$. By Pythagorean substitution, there are positive integers m, n such that $\gcd(m,n) = 1$ and

$$a = m^2 + n^2, \qquad 2x^2 = 2mn, \qquad y^2 = m^2 - n^2.$$

Since $x^2 = mn$ and $\gcd(m,n) = 1$, $m = p^2$ and $n = q^2$. So $y^2 = m^2 - n^2 = p^4 - q^4$. Thus (p,q,y) is a new solution of (1) with $p = \sqrt{m} \le x = \sqrt{r'/2} < r' \le a$, and so again we violate our extremal assumption.

Therefore, our assumption is false and there is no solution for (1) in positive integers. □

Now we apply the lemma to our problem. Solving the given equation for z^2, we find that the discriminant of the resulting quadratic is $x^4 + 6x^2y^2 + y^4$. Proceed indirectly again; suppose that $x^4 + 6x^2y^2 + y^4$ is a perfect square. Let $a = x^2 + y^2$ and $b = 2xy$. But now $a^2 + b^2$ and $a^2 - b^2$ are both perfect squares, and $a^4 - b^4$ must therefore be a perfect square as well! By our lemma, this is impossible, and we are finally done.

Problem 7

For n a given positive integer, find the smallest positive integer k for which there exist k 0-1 sequences of length $2n+2$, such that any other 0-1 sequence of length $2n+2$ matches one of the given ones in at least $n+2$ positions.

First Solution. The answer is $k = 4$. First we prove that $k \le 4$ by providing 4 sequences that will do the job. Let $a = 00\cdots0$, $b = 11\cdots1$, $c = 011\cdots1$, $d = 100\cdots0$ be four $2n+2$ sequences. For any $2n+2$ sequence s, let x_s and y_s be the respective numbers of 0's and 1's in s. If $x_s > y_s$, then we can match s with a; if $x_s < y_s$, we can match s with b; if $x_s = y_s = n+1$ and s starts with 0, we can match s with c; if $x_s = y_s = n+1$ and s starts with 1, we can match s with d. Therefore four sequences a, b, c, d can indeed do the job and $k \le 4$.

Now we prove $k \geq 4$. Suppose, on the contrary, there exist three sequences of length $2n+2$: $A = a_1a_2\cdots a_{2n+2}$, $B = b_1b_2\cdots b_{2n+2}$, $C = c_1c_2\cdots c_{2n+2}$ that will do the job. We claim that at least one of the following statements is true:

$$a_i = b_i \text{ or } c_i \qquad \text{for at least } n+1 \text{ values of } i;$$

$$b_i = c_i \text{ or } a_i \qquad \text{for at least } n+1 \text{ values of } i;$$

$$c_i = a_i \text{ or } b_i \qquad \text{for at least } n+1 \text{ values of } i;$$

In fact, if the first statement is not true, then there are $n+2$ values of i such that $a_i \neq b_i, c_i$. So for those i's, $b_i = c_i$ and the third statement is true.

WLOG, we suppose that the first statement is true. We are going to find a counterexample s that cannot match A, or B, or C for at least $n+2$ positions. Find exactly $n+1$ i's such that $a_i = b_i$ or c_i. For those i's, let $s_i = 1 - a_i$. Then s is different from A in at least $n+1$ positions. Moreover, s is different from B or C in these $n+1$ positions and there are still $n+1$ positions of s left to be assigned. So we can make s different from both B and C in at least $n+1$ positions as well, a contradiction. Thus our assumption is wrong and $k \geq 4$.

From the above we see that $k = 4$.

Second Solution. As in the first solution, we can find 4 sequences that ensure the required condition is true. Hence $k \leq 4$. If $k < 4$, then let $a_1^{(m)}a_2^{(m)}\ldots a_{2n+2}^{(m)}$, $m = 1, \ldots, k$, be k sequences that ensure the required condition is true. Let $S = \{(0,0), (0,1), (1,0), (1,1)\}$. For $i = 1, 2, \ldots, n+1$, $(a_{2i-1}^{(m)}, a_{2i}^{(m)}) \in S$. Since $k < 4$, there exists $(b_{2i-1}, b_{2i}) \in S$ and $(b_{2i-1}, b_{2i}) \neq (a_{2i-1}^{(m)}, a_{2i}^{(m)})$ for all m. Now for $m = 1, \ldots, k$, sequences $b_1b_2\ldots b_{2n+2}$ and $a_1^{(m)}a_2^{(m)}\ldots a_{2n+2}^{(m)}$ differ in at least $n+1$ places, a contradiction. Hence $k = 4$.

Problem 8

The polynomials $P_n(x,y)$ for $n = 1, 2, \ldots$ are defined by $P_1(x,y) = 1$ and

$$P_{n+1}(x,y) = (x+y-1)(y+1)P_n(x,y+2) + (y-y^2)P_n(x,y).$$

Prove that $P_n(x,y) = P_n(y,x)$ for all n and all x, y.

Solution. We induct on n. For $n = 1, 2$ the result is evident, as $P_1(x,y) = 1$ and $P_2(x,y) = xy + x + y - 1$. So we take $n > 1$ and suppose that the result is true for $P_{n-1}(x,y)$ and $P_n(x,y)$. We have

$$\begin{aligned}P_{n+1}(x,y) &= (x+y-1)(y+1)P_n(x,y+2)+(y-y^2)P_n(x,y)\\ &= (x+y-1)(y+1)P_n(y+2,x)+(y-y^2)P_n(y,x).\end{aligned}$$

Note that

$$\begin{aligned}&(x+y-1)(y+1)P_n(y+2,x)\\ &\quad= S_{n-1}(x,y)+(x+y-1)(y+1)(x-x^2)P_{n-1}(y+2,x),\end{aligned}$$

where $S_{n-1}(x,y) = [(x+y)^2-1](y+1)(x+1)P_{n-1}(y+2,x+2)$. Since $P_{n-1}(x,y) = P_{n-1}(y,x)$, $S_{n-1}(x,y) = S_{n-1}(y,x)$. We also notice that

$$\begin{aligned}&(y-y^2)P_n(y,x)\\ &\quad= (y-y^2)[(x+y-1)(x+1)P_{n-1}(y,x+2)+(x-x^2)P_{n-1}(y,x)]\\ &\quad= (y-y^2)(x+y-1)(x+1)P_{n-1}(y,x+2)+T_{n-1}(x,y),\end{aligned}$$

where $T_{n-1}(x,y) = (y-y^2)(x-x^2)P_{n-1}(y,x)$. Since $P_{n-1}(x,y) = P_{n-1}(y,x)$, $T_{n-1}(x,y) = T_{n-1}(y,x)$. Let

$$\begin{aligned}U_{n-1}(x,y) = (x+y-1)[&(y+1)(x-x^2)P_{n-1}(y+2,x)\\ &+(x+1)(y-y^2)P_{n-1}(y,x+2)].\end{aligned}$$

Since $P_{n-1}(x,y) = P_{n-1}(y,x)$, $U_{n-1}(x,y) = U_{n-1}(y,x)$. So

$$P_{n+1}(x,y) = S_{n-1}(x,y)+T_{n-1}(x,y)+U_{n-1}(x,y) = P_{n+1}(y,x).$$

This completes the induction. Therefore $P_n(x,y) = P_n(y,x)$ for all n and all x, y.

Problem 9

On the sides of a non-obtuse triangle ABC are constructed externally a square, a regular n-gon and a regular m-gon ($m, n > 5$) whose centers form an equilateral triangle. Prove that $m = n = 6$, and find the angles of triangle ABC.

Solution. The angles are $90°, 45°, 45°$. We prove the following lemma.

Lemma 2. *Let O be a point inside equilateral triangle XYZ. If*

$$\angle YOZ = x,\ \angle ZOX = y,\ \angle XOY = z.$$

then

$$\frac{OX}{\sin(x-60^\circ)} = \frac{OY}{\sin(y-60^\circ)} = \frac{OZ}{\sin(z-60^\circ)}.$$

Proof. Rotate triangle XYZ through an angle of 60° clockwise about Z. Let X' and O' be the images of X and O respectively. Then X is the image of Y, triangle $ZO'O$ is equilateral, and triangles $ZO'X$ and ZOY are congruent. Note that $x+y+z=360^\circ$. We have

$$\angle O'OX = \angle ZOX - 60^\circ = y - 60^\circ,$$
$$\angle XO'O = \angle XO'Z - 60^\circ = \angle YOZ - 60^\circ = x - 60^\circ,$$
$$\angle OXO' = 180^\circ - \angle O'OX - \angle XO'O = z - 60^\circ.$$

Note that $O'X = OY$. Applying the law of sines to triangle XOO' yields the desired result. □

Now we prove our main result. WLOG, suppose that the square is on AB, the m-gon is on BC, and the n-gon is on CA. Let O be the circumcenter of triangle ABC. WLOG, let the circumradius of triangle ABC be 1. Let X, Y, and Z be the respective centers of the square, the m-gon, and the n-gon. Let $\angle BAC = a$, $\angle CBA = b$, and $\angle ACB = c$. Then $OB = 1$, $\angle BOY = a$, and $\angle OYB = 180^\circ/m$. Let $\alpha = 180^\circ/m$. Applying the law of sines to triangle OBY,

$$OY = \frac{\sin(a+\alpha)}{\sin\alpha}.$$

Similarly, let $\angle ZOC = 180^\circ/n = \beta$. We have:

$$OX = \frac{\sin(c+45^\circ)}{\sin 45^\circ} = \sqrt{2}\sin(c+45^\circ), \quad OZ = \frac{\sin(b+\beta)}{\sin\beta}.$$

We apply the lemma to triangle XYZ and the point O; we have

$$\frac{OY}{\sin(b+c-60^\circ)} = \frac{OZ}{\sin(c+a-60^\circ)} = \frac{OX}{\sin(a+b-60^\circ)},$$

$$\frac{OY}{\sin(a+60^\circ)} = \frac{OZ}{\sin(b+60^\circ)} = \frac{OX}{\sin(c+60^\circ)},$$

$$\frac{\sin(a+\alpha)\csc\alpha}{\sin(a+60^\circ)} = \frac{\sin(b+\beta)\csc\beta}{\sin(b+60^\circ)} = \frac{\sqrt{2}\sin(c+45^\circ)}{\sin(c+60^\circ)}.$$

Note that function

$$f(x) = \frac{\sin(x+45^\circ)}{\sin(x+60^\circ)}$$

is increasing for $x \in [0°, 90°]$. In fact,

$$f'(x) = \frac{\cos(x+45°)\sin(x+60°) - \sin(x+45°)\cos(x+60°)}{\sin^2(x+60°)}$$
$$= \frac{\sin(15°)}{\sin^2(x+60°)} > 0.$$

Thus

$$\frac{\sin(a+\alpha)\csc\alpha}{\sin(a+60°)} = \frac{\sin(b+\beta)\csc\beta}{\sin(b+60°)} = \frac{\sqrt{2}\sin(c+45°)}{\sin(c+60°)}$$
$$= \sqrt{2}f(c) \leq \frac{\sqrt{2}\sin(90°+45°)}{\sin(90°+60°)} = 2.$$

Since the triangle is non-obtuse, at least two of its angles are between 45° and 90°. WLOG, let $45° \leq b\ (\leq 90°)$. Then

$$\frac{\sin(b+\beta)\csc\beta}{\sin(b+60°)} \leq 2 \iff \sin(b+\beta)\csc\beta \leq 2\sin(b+60°)$$
$$\iff \cos b + \sin b\cot\beta \leq \sin b + \sqrt{3}\cos b$$
$$\iff \cot\beta \leq 1 + (\sqrt{3}-1)\cot b$$
$$\implies \cot\beta \leq 1 + (\sqrt{3}-1)\cot 45° = \sqrt{3}.$$

Therefore, $\beta \geq 30°$. But since $n \geq 6$, $\beta = 180°/n \leq 30°$. So the equalities hold, $c = 90°$, and $b = a = 45°$, as claimed.

Problem 10

Let $a_1, \ldots, a_n$ be real numbers, not all zero. Prove that the equation

$$\sqrt{1+a_1x} + \cdots + \sqrt{1+a_nx} = n$$

has at most one nonzero real root.

Solution. Notice that $f_i(x) = \sqrt{1+a_ix}$ is concave. Hence $f(x) = \sqrt{1+a_1x} + \cdots + \sqrt{1+a_nx}$ is concave. Since $f'(x)$ exists, there can be at most one point on the curve $y = f(x)$ with derivative 0. Suppose there is more than one nonzero root. Since $x = 0$ is also a root, we have three real roots $x_1 < x_2 < x_3$. Applying the Mean-Value theorem to $f(x)$ on intervals $[x_1, x_2]$ and $[x_2, x_3]$, we can find two distinct points on the curve with derivative 0, a contradiction. Therefore, our assumption is wrong and there can be at most one nonzero real root for the equation $f(x) = n$.

Problem 11

Let m, n be natural numbers such that

$$A = \frac{(m+3)^n + 1}{3m}$$

is an integer. Prove that A is odd.

First Solution. Let p be a prime and a a positive integer. Let (a/p) be the Legendre symbol, i.e., (a/p) will have value 1 if a is a quadratic residue modulo p, -1 if a is a quadratic non-residue modulo p, and 0 if $p|a$. For odd primes p and q, we have LQR (Law of Quadratic Reciprocity):

$$(-1/p) = (-1)^{(p-1)/2}, \qquad (p/q)(q/p) = (-1)^{(p-1)(q-1)/4}.$$

If m is odd, then $(m+3)^n + 1$ is odd and A is odd. Now we suppose that m is even. Since A is an integer,

$$0 \equiv (m+3)^n + 1 \equiv m^n + 1 \pmod 3,$$

so $n = 2k+1$ is odd and $m \equiv -1 \pmod 3$. We consider the following cases.

(a) $m = 8m'$ for some positive integer m'. Then

$$(m+3)^n + 1 \equiv 3^{2k+1} + 1 \equiv 4 \pmod 8$$

and $3m \equiv 0 \pmod 8$. So A is not an integer.

(b) $m = 2m'$ for some odd positive integer m', i.e., $m \equiv 2 \pmod 4$. Then

$$(m+3)^n + 1 \equiv (2+3)^n + 1 \equiv 2 \pmod 4$$

and $3m \equiv 2 \pmod 4$. So A is odd.

(c) $m = 4m'$ for some odd positive integer m'. Since $m \equiv -1 \pmod 3$, there exists an odd prime p such that $p \equiv -1 \pmod 3$ and $p|m$. Since A is an integer,

$$0 \equiv (m+3)^n + 1 \equiv 3^{2k+1} + 1 \pmod m$$

and $3^{2k+1} \equiv -1 \pmod p$. Let a be a primitive root modulo p; let b be a positive integer such that $3 \equiv a^b \pmod p$. Thus $a^{(2k+1)b} \equiv -1 \pmod p$. Note that $(p/3) = (-1/3) = -1$. We consider the following cases.

(i) $p \equiv 1 \pmod 4$. From LQR, $(-1/p) = 1$, so

$$a^{2c} \equiv -1 \equiv a^{(2k+1)b} \pmod p$$

for some positive integer c. Therefore b is even and $(3/p) = 1$. Again, from LQR, we have

$$-1 = (3/p)(p/3) = (-1)^{(3-1)(p-1)/4} = 1,$$

a contradiction.

(ii) $p \equiv 3 \pmod 4$. From LQR, $(-1/p) = -1$, so

$$a^{2c+1} \equiv -1 \equiv a^{(2k+1)b} \pmod p$$

for some positive integer c. Therefore b is odd and $(3/p) = -1$. Again, from LQR, we have

$$1 = (3/p)(p/3) = (-1)^{(3-1)(p-1)/4} = -1,$$

a contradiction.

Thus for $m = 4m'$ and m' is odd, A is not an integer.

From the above, we see that if A is an integer, A is odd.

Second Solution. We prove by contradiction. Assume, on the contrary, that A is even. Then m is even. Since A is an integer, $0 \equiv (m+3)^n + 1 \equiv m^n + 1 \pmod 3$ yields $n = 2k+1$ and $m = 3t+2$. Let $m = 2^l m_1$, where $l \geq 1$ and m_1 is odd. In fact $l > 1$, as otherwise $m \equiv 2 \pmod 4$,

$$(m+3)^n + 1 \equiv (2+3)^n + 1 \equiv 2 \pmod 4$$

and

$$A = \frac{(m+3)^n + 1}{2m'}$$

is odd.

Since A is an integer, we have

$$0 \equiv (m+3)^n + 1 \equiv 3^{2k+1} + 1 \pmod m. \tag{2}$$

From (2), we have $2^l | (3^{2k+1} + 1)$. But

$$3^{2k+1} + 1 = 9^k \times 3 + 1 \equiv 4 \pmod 8,$$

so $l = 2$, $m = 4m_1$, and $m_1 \equiv m \equiv 2 \pmod 3$.

From (2), we also have $m_1 | (3^{2k+1} + 1)$, which implies that $m_1 | a^2 + 3$, where $a = 3^{k+1}$. Since

$$\gcd(m_1, 2) = \gcd(m_1, 3) = 1$$

and $m_1 \equiv 2 \pmod 3$, $m_1 = 6s+5$. Since $\gcd(m_1, a) = \gcd(m_1, 3) = 1$, Thue's lemma implies that there exist integers x and y such that $1 \leq x$,

$|y| \le \lfloor\sqrt{m_1}\rfloor$ and $m_1|ax+y$. Since $m_1 \equiv 2 \pmod 3$, m_1 is not a perfect square and $1 \le x, |y| < \sqrt{m_1}$. Now

$$m_1|a^2+3 \iff a^2+3 \equiv 0 \pmod{m_1},$$

$$m_1|ax+y \iff ax \equiv -y \pmod{m_1}$$

imply that $3x^2+y^2 \equiv 0 \pmod{m_1}$, i.e., $3x^2+y^2 = km_1$. But $3x^2+y^2 < 4m_1$ gives $k = 1, 2, 3$.

(a) If $k=1$, then $3x^2+y^2 = m_1 = 6s+5$ yields

$$y^2 \equiv 2 \pmod 3,$$

which is impossible.

(b) If $k=2$, then $3x^2+y^2 = 2m_1 = 12s+10$ yields

$$3x^2+y^2 \equiv 2 \pmod 4,$$

which is impossible.

(c) If $k=3$, then $3x^2+y^2 = 3m_1$ yields

$$y = 3y_1 \quad \text{and} \quad x^2+3y_1^2 = m_1.$$

We are back in the first case, which is impossible.

Thus our assumption is wrong and A is odd.

Problem 12

The sides and diagonals of a regular n-gon X are colored in k colors so that

(a) for any color a and any two vertices A, B of X, either AB is colored in a or AC and BC are colored in a for some other vertex C;

(b) the sides of any triangle with vertices among the vertices of X are colored in at most two colors.

Prove that $k \le 2$.

Solution. Observe that if we can reach a contradiction with $k=3$, then we will be done. We can reduce any $k>3$ case into a $k=3$ case by selecting three of the k colors, say red, blue, and green, to keep and recoloring in green every edge that is not already red, blue, or green. We present a lemma.

Lemma 3. *Suppose we have four vertices W, X, Y, and Z such that WX, XZ, and YZ are all colored differently, and WY is the same color as WX. Then WZ must be in the same color as WX and WY.*

Proof. By rule (b), each of triangles WXZ and WYZ cannot contain all three colors. Therefore, the only coloring that satisfies the rule is when WZ is the same color as WX and WY. □

We will demonstrate that this forces our n-gon to have infinitely many sides, a contradiction. Let us proceed by construction. By assumption, there must be at least one red edge; let that connect the vertices X and R_1. By rule (a), there must exist another vertex B_1 such that XB_1 and RB_1 are both blue. Also by rule (a), there must exist a vertex G_1 such that XG_1 and B_1G_1 are both green. Our lemma tells us that G_1R_1 must be green.

Observe that we constructed our vertices in the order X, R_1, B_1, and G_1. We will continue to construct vertices in the order R_i, B_i, G_i, R_{i+1}, B_{i+1}, $G_{i+1}\ldots$ such that whenever we construct a vertex, it must have edges of that color to all already-existing vertices. Yet we will also show that we will never be done (i.e., rule (a) will never be completely satisfied). An example:

Suppose that we have already constructed all of the vertices up to G_n. Now we add R_{n+1}. From our (inductive) assumption, for every vertex $Y \neq G_n, B_n$ already constructed, YB_n must be blue. Also, for every vertex $Y \neq G_n$ already constructed, YG_n must be green. Therefore, rule (a) is not satisfied for the vertices $\{B_n, G_n\}$ and the color red. Thus, there must exist some R_{n+1} such that $R_{n+1}B_n$ and $R_{n+1}G_n$ is red. But now observe that for every vertex P (other than R_{n+1}, B_n, and G_n) on our graph, we can apply our lemma with

$$W = R_{n+1}, X = B_n, Y = G_n, Z = P,$$

and $R_{n+1}P$ is also red. Similarly, we can observe that there must exist some B_{n+1} such that $B_{n+1}R_{n+1}$ and $B_{n+1}G_n$ is blue, and find that B_{n+1} must be connected to every existing vertex with a blue edge. Finally, there must also exist some G_{n+1} such that $G_{n+1}R_{n+1}$ and $G_{n+1}B_{n+1}$ is green; we also find that G_{n+1} must be connected to every existing vertex with a green edge.

We can continue this indefinitely, because after every "cycle," our G_n and B_n fail rule (a). Thus our polygon must have infinitely many vertices, and we are done.

Problem 13

Solve the following equation in natural numbers:

$$x^2 + y^2 = 1997(x - y).$$

Solution. The solutions are

$$(x, y) = (170, 145) \text{ or } (1827, 145).$$

We have

$$\begin{aligned} x^2 + y^2 &= 1997(x - y) \\ 2(x^2 + y^2) &= 2 \times 1997(x - y) \\ x^2 + y^2 + (x^2 + y^2 - 2 \times 1997(x - y)) &= 0 \\ (x + y)^2 + ((x - y)^2 - 2 \times 1997(x - y)) &= 0 \\ (x + y)^2 + (1997 - x + y)^2 &= 1997^2. \end{aligned}$$

Since x and y are positive integers, $0 < x + y < 1997$ and $0 < 1997 - x + y < 1997$. Thus the problem reduces to solving $a^2 + b^2 = 1997^2$ in positive integers. Since 1997 is a prime, $\gcd(a, b) = 1$. By Pythagorean substitution, there are positive integers $m > n$ such that $\gcd(m, n) = 1$ and

$$1997 = m^2 + n^2, \ a = 2mn, \ b = m^2 - n^2.$$

Since $m^2, n^2 \equiv 0, 1, -1 \pmod 5$ and $1997 \equiv 2 \pmod 5$, $m, n \equiv \pm 1 \pmod 5$. Since $m^2, n^2 \equiv 0, 1 \pmod 3$ and $1997 \equiv 2 \pmod 3$, $m, n \equiv \pm 1 \pmod 3$. Therefore $m, n \equiv 1, 4, 11, 14 \pmod{15}$. Since $m > n$, $1997/2 \le m^2 \le 1997$. Thus we only need to consider $m = 34, 41, 44$. The only solution is $(m, n) = (34, 29)$. Thus

$$(a, b) = (1972, 315),$$

which leads to our final solutions.

Problem 14

Let A_1 and B_1 be two points on the base AB of an isosceles triangle ABC ($\angle C > 60°$) such that $\angle A_1CB_1 = \angle ABC$. A circle externally tangent to the circumcircle of triangle A_1B_1C is tangent also to the rays CA and CB at points A_2 and B_2, respectively. Prove that $A_2B_2 = 2AB$.

Solution. Invert about C. Then A', B', A_1', and B_1' will all be concyclic with C. By inversive angles, $\angle A_1CB_1 = \angle BAC$. But $\angle A'B'C = \angle B'A'C = \angle B'A_1'C$; thus, the circumcenter of $A'B'C$ is equidistant from $A'C$, $B'C$, and $A_1'B_1'$. This means that the image of our externally tangent circle is the circle through the midpoints of $A'C$, $B'C$, and $A_1'B_1'$; thus A and B are the midpoints of CA_2 and CB_2, respectively, and we are done.

Problem 15

A square table $n \times n$ ($n \geq 2$) is filled with 0's and 1's so that any subset of n cells, no two of which lie in the same row or column, contains at least one 1. Prove that there exist i rows and j columns with $i+j \geq n+1$ whose intersection contains only 1's.

Solution. Consider the bipartite graph on the rows and columns (as vertices of the graph) such that row i is adjacent to column j if the ij-th entry in the matrix is zero. By assumption there is no matching of the vertices of this graph, so by the Marriage Theorem there exists a set S of i rows which are adjacent to a total of $k < i$ columns. Thus if T denotes the $j = n - k$ remaining columns, all entries in the intersection of S and T must be 1, and

$$i + j = i + (n - k) = n + (i - k) \geq n + 1.$$

Problem 16

Find all finite sets A of distinct nonnegative real numbers for which:

(a) the set A contains at least 4 numbers;

(b) for any 4 distinct numbers $a, b, c, d \in A$ the number $ab + cd \in A$.

Solution. We claim that $A = \{x, 1, 1/x, 0\}$ for some real number $x > 1$. We first prove the following lemma.

Lemma 4. *Let A be a set of distinct nonnegative real numbers such that*

(a) the set A contains at least 4 numbers;

(b) the value of each number in A is less than 1;

(c) for any 4 distinct numbers $a, b, c, d \in A$ the number $ab + cd \in A$.

Then A is an infinite set.

Proof. Suppose, on the contrary, that A is finite. Let $A = \{a_1, a_2, \ldots, a_n\}$, where $1 > a_1 > a_2 > \cdots > a_n \geq 0$ and $n \geq 4$. We first prove that $a_n \neq 0$. Assume for contradiction that $a_n = 0$. Then $a_1 a_n + a_{n-1} a_{n-2} \in A$. But $a_1 a_n = 0$ and $a_{n-1} a_{n-2} < a_{n-1}$, and hence there is an element of A in between a_{n-1} and a_n, which is impossible. Therefore if A is finite, $a_n \neq 0$. Now, let

$$a = a_1a_2 + a_3a_4, \quad b = a_1a_3 + a_2a_4, \quad c = a_1a_4 + a_2a_3.$$

Then $a, b, c \in A$ and $a > b > c$. If $c > a_4$, then $a = a_1$, $b = a_2$, $c = a_3$. From the first and the third equation, we have

$$a_1(1 - a_2) = a_3a_4,$$

$$a_3(1 - a_2) = a_1a_4.$$

Multiplying the equations gives

$$a_1a_3(1 - a_2)^2 = a_3a_1a_4^2.$$

Since $a_1, a_3 \neq 0$, we may now divide by a_1a_3 and take square roots to obtain $1 - a_2 = a_4$. But then $a_1 = a_3$, unless $a_4 = 0$. However, a_4 cannot equal zero, and we have reached a contradiction. So $c = a_1a_4 + a_2a_3 \leq a_4$, and $a_4(1 - a_1) \geq a_2a_3$. Thus $a_3 > a_4$ implies that

$$a_3(1 - a_1) \geq a_2a_3 \Longrightarrow 1 - a_1 \geq a_2 \Longrightarrow 1 \geq a_1 + a_2.$$

So $a_2 < 1/2$ and $a = a_1a_2 + a_3a_4 < a_1/2 + a_3/2 < a_1$. If $c = a_4$, then $a = a_2, b = a_3, c = a_4$. From the second equation and the third equation, we have

$$a_3(1 - a_1) = a_4a_2,$$

$$a_4(1 - a_1) = a_2a_3.$$

Using the same argument as before, we have $1 - a_1 = a_2 = 0$, a contradiction. Thus $c < a_4$ and a_5 exists. If we reapply the above argument to (a_2, a_3, a_4, a_5), we can prove the existence of a_6 and so on. Therefore A must be an infinite set. □

Now we prove our main result. We use the same notation. From the lemma, we know that in order for A to satisfy the given conditions, we must have $a_1 \geq 1$. Then $c > a_4$ and $a = a_1, b = a_2, c = a_3$. From the same argument used in the proof of the lemma, we have $1 - a_2 = a_4 = 0 \Longleftrightarrow a_2 = 1$ and $a_4 = 0$. Thus $b = a_1a_3 = a_2 = 1$. We have $A = \{a_1, 1, 1/a_1, 0\}$, as claimed.

Problem 17

Let ABC be an equilateral triangle and $n > 1$ be a positive integer. Denote by S the set of $n - 1$ lines which are parallel to AB and divide triangle

ABC into n parts of equal area, and by S' the set of $n-1$ lines which are parallel to AB and divide triangle ABC into n parts of equal perimeter. Prove that S and S' do not share a common element.

Solution. Suppose that the lines in S meet the triangle at A_1, B_1, A_2, B_2, …, A_{n-1}, B_{n-1} such that A_i are on AC, B_i are on BC, $A_iB_i \parallel AB$, and $CA_i < CA_j$ for $i < j$. Define A_i', B_i' for S' analogously . WLOG, let $AB = 1$.

Note that $[CA_1B_1]$, $[CA_2B_2]$, …, $[CA_{n-1}B_{n-1}]$, $[CAB]$ form an arithmetic sequence. Since all of these triangles are similar to each other,

$$(A_1B_1)^2,\ (A_2B_2)^2,\ \ldots,\ (A_{n-1}B_{n-1})^2,\ (A_nB_n)^2$$

form an arithmetic sequence. Since $[CA_1B_1] = [CAB]/n$,

$$A_1B_1 = \frac{1}{\sqrt{n}} \quad \text{and} \quad A_iB_i = \sqrt{\frac{i}{n}}.$$

On the other hand, $A_2'B_2' + A_1'A_2' + B_1'B_2' = CA_1' + CB_1'$. Let X be the point on $A_2'B_2'$ such that $A_1'A_2'XB_1'$ is a parallelogram. Then

$$A_2'B_2' = A_2'X + XB_2' = A_1'B_1' + XB_2' = CA_1' + XB_2'.$$

Hence $XB_2' + A_1'A_2' + B_1'B_2' = CB_1'$. But

$$A_1'A_2' = B_1'X = XB_2' = B_1'B_2',$$

so $3B_1'B_2' = CB_1'$. For any trapezoid $A_i'B_i'B_{i+2}'A_{i+2}'$, we can cut out a parallelogram to form an equilateral triangle and reapply the above argument to obtain $B_i'B_{i+1}' = 3B_{i+1}'B_{i+2}'$. Let $CB_1' = x$. Then

$$CB = 1 = x + \frac{x}{3} + \frac{x}{3^2} + \cdots + \frac{x}{3^{n-1}} = \frac{x(3^n-1)}{2\cdot 3^{n-1}}.$$

So $x = (2\cdot 3^{n-1})/(3^n - 1)$ and

$$A_i'B_i' = CB_i' = x + \frac{x}{3} + \cdots + \frac{x}{3^{i-1}} = x \cdot \frac{3^i - 1}{2\cdot 3^{i-1}} = \frac{3^{n-i}(3^i-1)}{3^n - 1}.$$

Since S and S' contain parallel lines, proving $S \cap S' = \emptyset$ is equivalent to proving that $A_iB_i \neq A_j'B_j'$ for any $i, j \in \{1, 2, \ldots, n-1\}$. Suppose, on the contrary, that $A_iB_i = A_j'B_j'$ for some $i, j \in \{1, 2, \ldots, n-1\}$. Then

$$\sqrt{\frac{i}{n}} = \frac{3^{n-j}(3^j-1)}{3^n-1}.$$

Let

$$j' = n - j,\ g = \gcd(3^j - 1, 3^n - 1),\ 3^n - 1 = mg,\ 3^{j-1} = lg.$$

Then 3 does not divide g, and m and l are relatively prime positive integers. We have

$$\frac{i}{n} = \left(\frac{3^{j'}l}{m}\right)^2 = \frac{3^{2j'}l^2}{m^2},$$

where the RHS is in simplest form. Note that

$$g|(3^n - 1) - (3^j - 1) = 3^n - 3^j = 3^j(3^{j'} - 1).$$

So $g|3^{j'} - 1$. Since $j + j' = n$, $\min\{j, j'\} \le n/2$ and $g \le 3^{n/2} - 1$. Therefore $m = (3^n - 1)/g \ge 3^{n/2} + 1$. But then $m^2 > 3^n > n$, a contradiction. Thus our assumption is false and $S \cap S' = \emptyset$.

Problem 18

Find all natural numbers n for which the polynomial $x^n + 64$ can be written as a product of two nonconstant polynomials with integer coefficients.

Solution. We claim that $n = 3k$ or $n = 4k$ for some positive integer k. Suppose that $x^n + 64 = f(x)g(x)$ for two nonconstant integer polynomials $f(x)$ and $g(x)$. Let $a = \sqrt[n]{64}$ and $\alpha_k = (2k+1)\pi/n$ for $k = 1, 2, \ldots, n$. Then

$$\begin{aligned} f(x)g(x) &= x^n + 64 \\ &= (x - a\operatorname{cis}\alpha_1)(x - a\operatorname{cis}\alpha_2)\cdots(x - a\operatorname{cis}\alpha_n). \end{aligned}$$

Since there is at most one real root for $x^n + 64$ (when n is odd), one of $f(x)$ or $g(x)$ has only complex roots. WLOG, let $f(x)$ be that polynomial. Let c be the constant term of $f(x)$ and d be the degree of $f(x)$. Since $g(x)$ is nonconstant, $d < n$. Since $f(x)$ has integer coefficients, all its roots can be paired as conjugates. So $2m = d < n$. Noticing $(a\operatorname{cis}\alpha_k)\overline{(a\operatorname{cis}\alpha_k)} = 64^{2/n}$, we have $c = 64^{2m/n} = (2^6)^{2m/n}$ an integer less than 64. Thus $2m/n$ can be reduced to a fraction with its denominator equal to 2, 3, or 6, so $3|n$ or $4|n$. Since

$$\begin{aligned} x^{3k} + 64 &= (x^k + 4)(x^{2k} - 4x^k + 16) \\ x^{4k} + 64 &= (x^{2k} + 4x^k + 8)(x^{2k} - 4x^k + 8), \end{aligned}$$

both $n = 3k$ and $n = 4k$ satisfy the condition of the problem. Therefore n is a multiple of 3 or 4, as claimed.

1.2 Canada

Problem 1

Determine the number of real solutions a of the equation

$$\left\lfloor \frac{a}{2} \right\rfloor + \left\lfloor \frac{a}{3} \right\rfloor + \left\lfloor \frac{a}{5} \right\rfloor = a.$$

Solution. There are 30 solutions. Since $\lfloor a/2 \rfloor$, $\lfloor a/3 \rfloor$, and $\lfloor a/5 \rfloor$ are integers, so is a. Now write $a = 30p + q$ for integers p and q, $0 \le q < 30$. Then

$$\begin{aligned} &\left\lfloor \frac{a}{2} \right\rfloor + \left\lfloor \frac{a}{3} \right\rfloor + \left\lfloor \frac{a}{5} \right\rfloor = a \\ &\iff 31p + \left\lfloor \frac{q}{2} \right\rfloor + \left\lfloor \frac{q}{3} \right\rfloor + \left\lfloor \frac{q}{5} \right\rfloor = 30p + q \\ &\iff p = q - \left\lfloor \frac{q}{2} \right\rfloor - \left\lfloor \frac{q}{3} \right\rfloor - \left\lfloor \frac{q}{5} \right\rfloor . \end{aligned}$$

Thus, for each value of q, there is exactly one value of p (and one value of a) satisfying the equation. Since q can equal any of thirty values, there are exactly 30 solutions, as claimed.

Problem 2

Find all real numbers x such that

$$x = \left(x - \frac{1}{x} \right)^{1/2} + \left(1 - \frac{1}{x} \right)^{1/2}.$$

First Solution. $x = (1+\sqrt{5})/2$ is the only solution. Isolating $\sqrt{1 - 1/x}$ and squaring, we obtain

$$x^2 - 2\sqrt{x^3 - x} + x - 1/x = 1 - 1/x.$$

Isolating $2\sqrt{x^3 - x}$ and squaring again, we obtain

$$\begin{aligned} x^4 + 2x^3 - x^2 - 2x + 1 &= 4x^3 - 4x \\ x^4 - 2x^3 - x^2 + 2x + 1 &= 0 \\ (x-2)(x-1)x(x+1) + 1 &= 0. \end{aligned}$$

This equation is symmetric about $x = 1/2$, so we make the substitution $u = x - 1/2$, $x = u + 1/2$ to get

$$u^4 - \frac{5}{2}u^2 + \frac{25}{16} = (u^2 - 5/4)^2 = 0.$$

Therefore, $u = \pm\sqrt{5}/2$ and $x = (1 \pm \sqrt{5})/2$. Checking these in our original equation, only $x = (1 + \sqrt{5})/2$ is a valid solution.

Second Solution. From the original equation, we have $x > 0$ and $1 > 1/x$, thus $x > 1$. Now isolating $\sqrt{1 - 1/x}$ and squaring as in the first solution, we obtain

$$\begin{aligned}(x^2 - 1) - 2\sqrt{x(x^2 - 1)} + x &= 0\\ (\sqrt{x^2 - 1} - \sqrt{x})^2 &= 0.\end{aligned}$$

The last equation can be easily solved to obtain the solution.

Problem 3

Let n be a natural number such that $n \geq 2$. Show that

$$\frac{1}{n+1}\left(1 + \frac{1}{3} + \cdots + \frac{1}{2n-1}\right) > \frac{1}{n}\left(\frac{1}{2} + \frac{1}{4} + \cdots + \frac{1}{2n}\right).$$

Solution. We prove that

$$n\left(1 + \frac{1}{3} + \cdots + \frac{1}{2n-1}\right) > (n+1)\left(\frac{1}{2} + \frac{1}{4} + \cdots + \frac{1}{2n}\right).$$

by induction. For $n = 2$, $8/3 > 9/4$. Assume the claim is true for $n = k \geq 2$, i.e., we have

$$k\left(1 + \frac{1}{3} + \cdots + \frac{1}{2k-1}\right) > (k+1)\left(\frac{1}{2} + \frac{1}{4} + \cdots + \frac{1}{2k}\right).$$

Now let $n = k + 1$. Note that

$$\begin{aligned}&\left(1 + \frac{1}{3} + \cdots + \frac{1}{2k-1}\right) + \frac{k+1}{2k+1}\\ &\quad = \left(\frac{1}{2} + \frac{1}{3} + \cdots + \frac{1}{2k-1}\right) + \frac{1}{2} + \frac{k+1}{2k+1}\\ &\quad > \left(\frac{1}{2} + \frac{1}{4} + \cdots + \frac{1}{2k}\right) + \frac{1}{2} + \frac{k+1}{2k+1}\\ &\quad > \left(\frac{1}{2} + \frac{1}{4} + \cdots + \frac{1}{2k}\right) + \frac{k+1}{2k+2} + \frac{1}{2k+1}\\ &\quad > \left(\frac{1}{2} + \frac{1}{4} + \cdots + \frac{1}{2k}\right) + \frac{k+2}{2k+2}.\end{aligned}$$

Induction is complete as

$$\begin{aligned}(k+1)&\left(1+\frac{1}{3}+\cdots+\frac{1}{2k-1}+\frac{1}{2k+1}\right)\\ &=k\left(1+\frac{1}{3}+\cdots+\frac{1}{2k-1}\right)+\left(1+\frac{1}{3}+\cdots+\frac{1}{2k-1}\right)+\frac{k+1}{2k+1}\\ &>k\left(1+\frac{1}{3}+\cdots+\frac{1}{2k-1}\right)+\left(\frac{1}{2}+\frac{1}{4}+\cdots+\frac{1}{2k}\right)+\frac{k+2}{2k+2}\\ &>(k+2)\left(\frac{1}{2}+\frac{1}{4}+\cdots+\frac{1}{2k+2}\right).\end{aligned}$$

Problem 4

Let ABC be a triangle with $\angle BAC = 40^\circ$ and $\angle ABC = 60^\circ$. Let D and E be the points lying on the sides AC and AB, respectively, such that $\angle CBD = 40^\circ$ and $\angle BCE = 70^\circ$. Let BD and CE meet at F. Show that the line AF is perpendicular to the line BC.

Solution. Note that $\angle ABD = 20^\circ$, $\angle BCA = 80^\circ$, and $\angle ACE = 10^\circ$. Let G be the foot of the altitude from A to BC. Then $\angle BAG = 90^\circ - \angle ABC = 30^\circ$ and $\angle CAG = 90^\circ - \angle BCA = 10^\circ$. Now,

$$\begin{aligned}\frac{\sin\angle BAG\sin\angle ACE\sin\angle CBD}{\sin\angle CAG\sin\angle BCE\sin\angle ABD}&=\frac{\sin 30^\circ\sin 10^\circ\sin 40^\circ}{\sin 10^\circ\sin 70^\circ\sin 20^\circ}\\ &=\frac{(1/2)(\sin 10^\circ)(2\sin 20^\circ\cos 20^\circ)}{\sin 10^\circ\cos 20^\circ\sin 20^\circ}\\ &=1.\end{aligned}$$

Then by the trigonometric form of Ceva's Theorem, AG, BD, and CE are concurrent. Therefore, F lies on AG so AF is perpendicular to the line BC, as desired.

Problem 5

Let m be a positive integer. Define the sequence $\{a_n\}_{n\geq 0}$ by $a_0 = 0$, $a_1 = m$ and $a_{n+1} = m^2a_n - a_{n-1}$ for $n \geq 1$. Prove that an ordered pair (a, b) of nonnegative integers, with $a \leq b$, is a solution of the equation

$$\frac{a^2+b^2}{ab+1}=m^2$$

if and only if $(a,b) = (a_n, a_{n+1})$ for some $n \geq 0$.

Solution. The "if" direction of the claim is easily proven by induction on n; we prove the "only if" direction by contradiction. Suppose, on the contrary, that there exist pairs satisfying the equation but not of the described form; let (a, b) be such a pair with minimal sum $a+b$. We claim that $(c, a) = (m^2a - b, a)$ is another such a pair but with smaller sum $c + a$, which leads to a contradiction.

(a) $a = 0$. Then $(a, b) = (0, m) = (a_0, a_1)$, a contradiction.

(b) $a = m$. Then $(a, b) = (m, m^3) = (a_1, a_2)$, a contradiction.

(c) $a = 1$. Then $b \geq a = 1$ and $(b+1)|(b^2+1)$; but $(b+1)|(b^2-1)$, thus $(b+1)|[(b^2+1)-(b^2-1)] = 2$. We have $b = 1$, thus $m = 1$ and $(a, b) = (1, 1) = (a_1, a_2)$, a contradiction.

(d) $2 \leq a < m$. Rewrite $(a^2+b^2)/(ab+1) = m^2$ as

$$b^2 - m^2ab + a^2 - m^2 = 0,$$

we know that $t = b$ is a root of the quadratic equation

$$t^2 - m^2at + a^2 - m^2 = 0. \tag{1}$$

Thus $m^4a^2 + 4m^2 - 4a^2$ the discriminant of the equation must be a perfect square. But

$$\begin{aligned}(m^2a+1)^2 &= m^4a^2 + 2m^2a + 1 \\ &> m^4a^2 + 4m^2 - 4a^2 > (m^2a)^2\end{aligned}$$

for $2 \leq a < m$. So the discriminant cannot be a perfect square, a contradiction.

(e) $a > m$. Again $t = b$ is a root of (1). It is easy to check that $t = m^2a - b = c$ also satisfies the equation. We have $bc = a^2 - m^2 > 0$; since $b \geq 0$, $c > 0$. Since $a > 0$ and $c > 0$, $ac + 1 > 0$, we have

$$\frac{c^2+a^2}{ca+1} = m^2.$$

Since $c > 0$, $b \geq a$ and $bc = a^2 - m^2 < a^2$, we have $c < a$. Thus (c, a) is a valid pair. Also, it cannot be of the form (a_n, a_{n+1}) or else

$$(a, b) = (a_{n+1}, m^2a_{n+1} - a_n) = (a_{n+1}, a_{n+2}).$$

But then, $c + a < a + a \leq b + a$, as desired.

From the above, we see that our assumption is false. Therefore every pair satisfying the original equation must be of the described form.

1.3 China

Problem 1

Let ABC be a non-obtuse triangle such that $AB > AC$ and $\angle B = 45^\circ$. Let O and I denote the circumcenter and incenter of $\triangle ABC$ respectively. Suppose that $\sqrt{2}OI = AB - AC$. Determine all the possible values of $\sin\angle BAC$.

First Solution. Let

$$a = BC,\quad b{=}CA,\quad c = AB,$$

$$\alpha = \angle CAB,\quad \beta = \angle ABC,\quad \gamma = \angle BCA,$$

and let R and r be the circumradius and inradius of ABC, respectively. Applying law of sines to ABC, we have

$$a = 2R\sin\alpha,\ b = 2R\sin\beta,\ c = 2R\sin\gamma.$$

Since $\beta = 45^\circ, \sin\beta = \sqrt{2}/2, \tan(\beta/2) = (\sqrt{2}-1)$, and

$$\sin\gamma = \sin(135^\circ - \alpha) = \frac{\sqrt{2}(\sin\alpha + \cos\alpha)}{2}. \tag{1}$$

Thus

$$r = \frac{(c+a-b)}{2}\tan(\beta/2) = R(\sqrt{2}-1)(\sin\alpha + \sin\gamma - \sin\beta).$$

From Euler's formula $OI^2 = R(R-2r)$, we have

$$OI^2 = R^2(1 - 2(\sin\alpha + \sin\gamma - \sin\beta)(\sqrt{2}-1)). \tag{2}$$

Since $\sqrt{2}OI = AB - AC$,

$$OI^2 = (c-b)^2/2 = 2R^2(\sin\gamma - \sin\beta)^2.$$

From (1) and (2), we obtain

$$\begin{aligned}
&2(\sin\gamma - \sin\beta)^2 = (1 - 2(\sin\alpha + \sin\gamma - \sin\beta)(\sqrt{2}-1))\\
&\iff 1 - 2(\sin\gamma - \sin\beta)^2 = 2(\sin\alpha + \sin\gamma - \sin\beta)(\sqrt{2}-1)\\
&\iff 1 - 2\sin^2\gamma + 2\sqrt{2}\sin\gamma - 1\\
&\qquad = 2(\sin\alpha + \sin\gamma)(\sqrt{2}-1) - (2-\sqrt{2})\\
&\iff -(\sin\alpha + \cos\alpha)^2 + 2(\sin\alpha + \cos\alpha)\\
&\qquad = (2\sqrt{2}-2)\sin\alpha + (2-\sqrt{2})(\sin\alpha + \cos\alpha) - (2-\sqrt{2})
\end{aligned}$$

$$\begin{aligned}
&\iff -1 - 2\sin\alpha\cos\alpha \\
&\qquad = (\sqrt{2}-2)\sin\alpha - \sqrt{2}\cos\alpha - (2-\sqrt{2}) \\
&\iff 2\sin\alpha\cos\alpha - (2-\sqrt{2})\sin\alpha - \sqrt{2}\cos\alpha + (\sqrt{2}-1) = 0 \\
&\iff (\sqrt{2}\sin\alpha - 1)(\sqrt{2}\cos\alpha - \sqrt{2}+1) = 0.
\end{aligned}$$

Thus $\sin\alpha = \sqrt{2}/2$ or $\cos\alpha = 1 - \sqrt{2}/2$,

$$\sin\alpha = \sqrt{1-\cos^2\alpha} = \frac{\sqrt{4\sqrt{2}-2}}{2}.$$

Second Solution. Let I_c, I_a, I_b be the feet of perpendiculars from I to AB, BC, CA, respectively. Let D be the foot of the perpendicular from O to BC. Thus OD is the perpendicular bisector of BC and $BD = CD$. From equal tangents, we have $AI_c = AI_b$, $BI_a = BI_c$, $CI_a = CI_b$. We have

$$\begin{aligned}
\sqrt{2}OI &= c - b \\
&= (AI_c + I_cB) - (AI_b + I_bC) = I_cB - I_bC \\
&= BI_a - I_aC.
\end{aligned}$$

Since $c > b$, D is on BI_a. We have $BI_a = BD + DI_a$, $I_aC = CD - DI_a$. So $\sqrt{2}OI = 2DI_a$, i.e., $OI = \sqrt{2}DI_a$. Thus line OI and line DI_a form a 45° angle, which implies that either $OI \perp AB$ or $OI \parallel AB$.

(a) $OI \perp AB$. Then OI is the perpendicular bisector of AB. Thus $AC = BC$, $\alpha = \beta = 45^\circ$, and $\sin\alpha = \sqrt{2}/2$.

(b) $OI \parallel AB$. Let E be the foot of the perpendicular from O to AB. So $\angle AOE = \angle C = \gamma$,

$$R\cos\angle AOE = R\cos\gamma = OE = II_c = r.$$

Since

$$r = 4R\sin\frac{\alpha}{2}\sin\frac{\beta}{2}\sin\frac{\gamma}{2},$$

we have

$$\begin{aligned}
\cos\gamma &= 4\sin\frac{\alpha}{2}\sin\frac{\beta}{2}\sin\frac{\gamma}{2} = 2\sin\frac{\beta}{2}\left(2\sin\frac{\alpha}{2}\sin\frac{\gamma}{2}\right) \\
&= 2\sin\frac{\beta}{2}\left(-\cos\frac{\alpha+\gamma}{2} + \cos\frac{\alpha-\gamma}{2}\right) \\
&= 2\sin\frac{\beta}{2}\left(-\sin\frac{\beta}{2} + \cos\frac{\alpha-\gamma}{2}\right)
\end{aligned}$$

$$= -2\sin^2\frac{\beta}{2} + 2\sin\frac{\beta}{2}\cos\frac{\alpha-\gamma}{2}$$

$$= \cos\beta - 1 + \sin\frac{\alpha+\beta-\gamma}{2} + \sin\frac{\beta+\gamma-\alpha}{2}$$

$$= \cos\beta - 1 + \sin(90^\circ - \gamma) + \sin(90^\circ - \alpha)$$

$$= \cos\beta - 1 + \cos\gamma + \cos\alpha,$$

which implies that $\cos\alpha = 1 - \cos\beta = 1 - \sqrt{2}/2$ and

$$\sin\alpha = \sqrt{1-\cos^2\alpha} = \frac{\sqrt{4\sqrt{2}-2}}{2}.$$

Problem 2

Let n, $n > 1$, be a positive integer. Determine if there exist $2n$ distinct positive integers $a_1, a_2, \cdots, a_n, b_1, b_2, \cdots, b_n$ such that

(a) $a_1 + a_2 + \cdots + a_n = b_1 + b_2 + \cdots + b_n$;

(b) $n - 1 > \sum_{i=1}^{n} \frac{a_i - b_i}{a_i + b_i} > n - 1 - \frac{1}{1998}$.

Solution. The answer is yes. First we prove the following lemma.

Lemma 5. *If $a_1, a_2, \cdots, a_n, b_1, b_2, \cdots, b_n$ are $2n$ distinct positive integers such that $a_1 + a_2 + \cdots + a_n = b_1 + b_2 + \cdots + b_n$, then*

$$n - 1 > \sum_{i=1}^{n} \frac{a_i - b_i}{a_i + b_i}.$$

Proof. From the given conditions, we know that there exists i_1, $1 \le i_1 \le n$, such that $b_{i_1} > a_{i_1}$. So $2b_{i_1}/(a_{i_1} + b_{i_1}) > 1$ and

$$\sum_{i=1}^{n} \frac{a_i - b_i}{a_i + b_i} = \sum_{i=1}^{n}\left(1 - \frac{2b_i}{a_i + b_i}\right) = n - \sum_{i=1}^{n} \frac{2b_i}{a_i + b_i} < n - 1. \quad (3)$$

as claimed. □

Let N be a positive integer. For $i = 1, 2, \ldots, n-1$, let $a_i = N(2i-1)$, $b_i = 2i$. From (a), we have

$$\sum_{i=1}^{n} a_i = -(n-1)N + 2N \cdot \frac{(n-1)n}{2} + a_n$$

$$= 2 \cdot \frac{(n-1)n}{2} + b_n = \sum_{i=1}^{n} b_i.$$

Thus $b_n = a_n + (n-1)[N(n-1)-n]$ with a_n still to be determined. From the lemma, we only need to prove the second inequality in (b). But (3) implies that

$$n-1 > \sum_{i=1}^{n} \frac{a_i - b_i}{a_i + b_i}$$

$$= n - \frac{2a_n + 2(n-1)[N(n-1)-n]}{2a_n + (n-1)[N(n-1)-n]} - \sum_{i=1}^{n-1} \frac{4i}{2i + N(2i-1)}.$$

This quantity approaches $n-1$ as

$$N \longrightarrow +\infty \quad \text{and} \quad \frac{a_n}{N} \longrightarrow +\infty,$$

(for example, let $a_n = N^2$). Thus for for any $\epsilon > 0$, (here $\epsilon = 1/1998$), we can find large enough N and a_n, such that a_i, b_i are all distinct, $\sum_{i=1}^{n} a_i = \sum_{i=1}^{n} b_i$, and

$$n-1 > \sum_{i=1}^{n} \frac{a_i - b_i}{a_i + b_i} > n - 1 - \epsilon.$$

Problem 3

For a set U, let $|U|$ denote the number of elements in U. A set U is called *primitive* if all of its elements are positive integers and there is an element $u \in U$ such that $\gcd(u, u') = 1$ for all $u' \in U$, $u' \neq u$. A set U is called *non-primitive* if all of its elements are positive integers and there is an element $u \in U$ such that $\gcd(u, u') > 1$ for all $u' \in U$. Let set $S = \{1, 2, \ldots, 98\}$. Determine the minimum value of positive integer n such that, for any subset $T \subset S$ with $|T| = n$, it always possible to find a subset $T_{10} \subset T$ such that $|T_{10}| = 10$ and for any partition A and B of T_{10} with $|A| = |B| = 5$, one of them is primitive and the other is non-primitive.

Solution. The answer is $n = 50$. Let $T \subset S$. We say T is good if there exists $T_{10} \subset T$ satisfying conditions (a) and (b). Now let $|T| = 50$. We claim that all such T are good.

Let $E(T)$ be the number of even elements in T and $O(T)$ be the number of odd elements in T. We have $E(T) + O(T) = |T| = 50$. For an odd number $x \in S$, let $f(x)$ denote the number of even numbers y in S such that $\gcd(x, y) > 1$. We establish the following facts and lemma.

(1) $1 \in S$ and the first few odd primes in S are 3, 5, 7, 11, 13, 17, 19, 23, 29, 31, 37, 41, 43, 47, 53, 59.

(2) There are 16 odd numbers, 3, 9, 15, ..., 93, in S with smallest prime divisor 3.

(3) There are 7 odd numbers, 5, 25, 35, 55, 65, 85, 95, in S with smallest prime divisor 5.

(4) There are 4 odd numbers, 7, 49, 77, 91 in S with smallest prime divisor 7.

(5) There is 1 odd number in S with smallest prime divisor p, for all primes $11 \leq p \leq 97$.

Lemma 6. *If there is an odd number $x \in T$ such that*

$$f(x) \leq E(T) - 9,$$

then T is good.

Proof. Since $E(T) - 9 \geq f(x)$, there are 9 even numbers $t_1, \ldots, t_9 \in T$ such that $\gcd(x, t_i) = 1$. We define $T_{10} = \{x, t_1, \ldots, t_9\}$. Therefore T is good. □

Since there are only 49 even elements in S and $|T| = 50$, there is always an odd element $x \in T$. Let p be the smallest prime factor of x. Now we consider the following cases according to the values of $E(T)$.

(a) $E(T) \geq 31$. Since x can have at most two distinct prime factors,

$$f(x) \leq f(15) = \left\lfloor \frac{98}{6} \right\rfloor + \left\lfloor \frac{98}{10} \right\rfloor - \left\lfloor \frac{98}{30} \right\rfloor = 22.$$

Therefore $E(T) - 9 \geq 22 \geq f(x)$. From the lemma, T is good.

(b) $30 \geq E(T) \geq 21$. Then $O(T) = 50 - E(T) \geq 20$. Since

$$1 + 16 = 17 < 20,$$

from facts (1) and (2), we can further suppose that $p \geq 5$ and $x \neq 35, 55$. Thus

$$f(x) \leq f(65) = \left\lfloor \frac{98}{10} \right\rfloor + \left\lfloor \frac{98}{26} \right\rfloor = 12.$$

Therefore $E(T) - 9 \geq 12 \geq f(x)$. From the lemma, T is good.

(c) $20 \geq E(T) \geq 12$. Then $O(T) = 50 - E(T) \geq 30$. Since

$$1 + 16 + 7 + 4 = 28 = 30 - 2 \leq O(T) - 2,$$

from facts (1) to (5), we can further suppose that $p \geq 13$. Thus

$$f(x) \leq f(13) = \left\lfloor \frac{98}{26} \right\rfloor = 3 = 12 - 9 \leq E(T) - 9.$$

By the lemma, T is good.

(d) $11 \geq E(T) \geq 10$. Then $O(T) = 50 - E(T) \geq 39$. Since

$$1 + 16 + 7 + 4 = 28 = 39 - 11 \leq O(T) - 11,$$

from facts (1) to (5), we can further suppose that $x = p \geq 47$. Thus

$$f(x) \leq f(47) = \left\lfloor \frac{98}{94} \right\rfloor = 1 = 10 - 9 \leq E(T) - 9.$$

By the lemma, T is good.

(e) $E(T) = 9$. Then $O(T) = 41$. Since

$$1 + 16 + 7 + 4 = 28 = 41 - 13 = O(T) - 13,$$

from facts (1) to (5), we can further suppose that $x = p \geq 59$. Thus

$$f(x) \leq f(59) = 0 = E(T) - 9.$$

By the lemma, T is good.

(f) $E(T) \leq 8$. Then $O(T) = 50 - E(T) \geq 42$. Since

$$1 + 16 + 7 + 4 = 28 = 42 - 14 = O(T) - 14,$$

from facts (1) to (5), we can further suppose that $x = p \geq 61$. Since $O(T) \leq 49$, there are at most 7 odd numbers not in T. Among $\lfloor 98/6 \rfloor + 1 = 16$ odd numbers in S that are divisible by 3, there are at least $16 - 7 = 9$ of them in T. Let $t_1, t_2, \ldots, t_9$ be 9 such numbers. Then we can let $T_{10} = \{x, t_1, \ldots, t_9\}$ and thus T is good.

So all T with $|T| = 50$ are good. But the subset of 49 even numbers is not good. Thus $n = 50$.

Problem 4

Determine all the positive integers $n \geq 3$, such that 2^{2000} is divisible by

$$1 + \binom{n}{1} + \binom{n}{2} + \binom{n}{3}.$$

Solution. The solutions are $n = 3, 7, 23$. Since 2 is a prime, $1 + \binom{n}{1} + \binom{n}{2} + \binom{n}{3} = 2^k$ for some positive integer $k \leq 2000$. We have $1 + \binom{n}{1} + \binom{n}{2} + \binom{n}{3} = (n+1)(n^2 - n + 6)/6$, i.e., $(n+1)(n^2 - n + 6) =$

$3 \times 2^{k+1}$. Let $m = n+1$, then $m \geq 4$ and $m(m^2 - 3m + 8) = 3 \times 2^{k+1}$. We consider the following two cases.

(a) $m = 2^s$. Since $m \geq 4$, $s \geq 2$. We have $2^{2s} - 3 \times 2^s + 8 = m^2 - 3m + 8 = 3 \times 2^t$ for some positive integer t. If $s \geq 4$, then $8 \equiv 3 \times 2^t \pmod{16} \Longrightarrow 2^t = 8 \Longrightarrow m^2 - 3m + 8 = 24 \Longrightarrow m(m-3) = 16$, which is impossible. Thus either $s = 3, m = 8, t = 4, n = 7$, or $s = 2, m = 4, t = 2, n = 3$.

(b) $m = 3 \times 2^u$. Since $m \geq 4$, $m > 4$ and $u \geq 1$. We have $9 \times 2^{2u} - 9 \times 2^u + 8 = m^2 - 3m + 8 = 2^v$ for some positive integer v. It is easy to check that there is no solution for v when $u = 1, 2$. If $u \geq 4$, we have $8 \equiv 2^v \pmod{16} \Longrightarrow v = 3$ and $m(m-3) = 0$, which is impossible. So $u = 3$, $m = 3 \times 2^3 = 24$, $v = 9$, $n = 23$.

Problem 5

Let D be a point inside an acute triangle ABC such that

$$DA \cdot DB \cdot AB + DB \cdot DC \cdot BC + DC \cdot DA \cdot CA = AB \cdot BC \cdot CA.$$

Determine the geometric position of D.

Lemma 7. *Let D be a point inside an acute triangle ABC. We have*

$$DA \cdot DB \cdot AB + DB \cdot DC \cdot BC + DC \cdot DA \cdot CA \geq AB \cdot BC \cdot CA; \quad (4)$$

equality holds if and only if D is the orthocenter of ABC.

It is clear that the lemma contains our main result. We are going to prove the lemma in two ways.

First Solution. Let E and F be points such that $BCDE$ and $BCAF$ are both parallelograms. Thus $EDAF$ is also a parallelogram. We have

$$AF = ED = BC,\ EF = AD,\ EB = CD,\ BF = AC.$$

Applying Ptolemy's theorem to quadrilaterals $ABEF$ and $AEBD$, we have

$$\begin{aligned} AB \cdot AD + BC \cdot CD &= AB \cdot EF + AF \cdot BE \\ &\geq AE \cdot BF = AE \cdot AC; \\ BD \cdot AE + AD \cdot CD &= BD \cdot AE + AD \cdot BE \\ &\geq AB \cdot ED = AB \cdot BC. \end{aligned}$$

Now we have

$$\begin{aligned}DA\cdot DB\cdot AB+DB\cdot DC\cdot BC+DC\cdot DA\cdot CA\\&=DB(AB\cdot AD+BC\cdot CD)+DC\cdot DA\cdot CA\\&\geq DB\cdot AE\cdot AC+DC\cdot DA\cdot CA\\&\geq AC(BD\cdot AE+AD\cdot CD)\\&\geq AC\cdot AB\cdot BC.\end{aligned}$$

Equality holds if and only if both $ABEF$ and $AEBD$ are cyclic, which implies that $AFEBD$ and $AFED$ are cyclic. Since $AFED$ is a parallelogram, $AFED$ is a rectangle and $AD\perp ED$. Since $BCDE$ is a parallelogram, we have $ED \parallel BC$and $AD\perp BC$. Since $AEBD$ is cyclic, $\angle ABE=\angle ADE$, which implies that $BE\perp AB$. Since $BCDE$ is a parallelogram, we have $CD\parallel EB$ and $CD\perp AB$. Thus D is the orthocenter of ABC.

Second Solution. Let D be the origin of the complex plane and let the complex coordinates of A, B, C be u, v, w, respectively. We rewrite (4) as

$$|uv(u-v)|+|vw(v-w)|+|wu(w-u)|\geq|(u-v)(v-w)(w-u)|. \quad (5)$$

But it is easy to check that

$$uv(u-v)+vw(v-w)+wu(w-u)=-(u-v)(v-w)(w-u), \quad (6)$$

which implies (5) and thus (4). Now we only need to determine when the equality holds. Let

$$z_1=\frac{uv}{(u-w)(v-w)},\quad z_2=\frac{vw}{(v-u)(w-u)},\quad z_3=\frac{wu}{(w-v)(u-v)}.$$

We can rewrite (5) and (6) as

$$|z_1|+|z_2|+|z_3|\geq 1$$

$$z_1+z_2+z_3=1.$$

Equality holds if and only if z_1, z_2, z_3 are all positive real numbers.

Suppose that z_1, z_2, z_3 are all positive real numbers. Since

$$-\frac{z_1z_2}{z_3}=\left(\frac{v}{w-u}\right)^2,\quad -\frac{z_2z_3}{z_1}=\left(\frac{w}{u-v}\right)^2,\quad -\frac{z_3z_1}{z_2}=\left(\frac{u}{v-w}\right)^2,$$

we know that $u/(v-w)$ and $v/(w-u)$ are pure imaginary numbers; thus $AD \perp BC$ and $BD \perp AC$ and D is the orthocenter of ABC.

Suppose that D is the orthocenter of the triangle ABC. Since the triangle is acute, D is inside the triangle. Therefore there are some positive numbers r_1, r_2, r_3 such that

$$\frac{u}{v-w} = -r_1 i, \qquad \frac{v}{w-u} = -r_2 i, \qquad \frac{w}{u-v} = -r_3 i.$$

Thus z_1, z_2, z_3 are all positive real numbers.

From the above, we know that the equality in (4) holds if and only if D is the orthocenter of ABC.

Problem 6

Let n, $n \geq 2$, be a positive integer. Let $x_1, x_2, \ldots, x_n$ be real numbers such that

$$\sum_{i=1}^{n} x_i^2 + \sum_{i=1}^{n-1} x_i x_{i+1} = 1.$$

Given an integer k, $1 \leq k \leq n$, determine the maximum value of $|x_k|$.

First Solution. The answer is

$$|x_k|_{\max} = \sqrt{\frac{2k(n+1-k)}{n+1}}.$$

We have

$$x_1^2 + (x_1+x_2)^2 + (x_2+x_3)^2 + \cdots + (x_{n-1}+x_n)^2 + x_n^2 = 2.$$

From the RMS-AM inequality, we have

$$\begin{aligned}
&\sqrt{\frac{x_1^2 + (x_1+x_2)^2 + \cdots + (x_{k-1}+x_k)^2}{k}} \\
&\qquad \geq \frac{|x_1| + |x_1+x_2| + \cdots + |x_{k-1}+x_k|}{k} \\
&\qquad \geq \frac{|x_1 - (x_1+x_2) + \cdots + (-1)^{k-1}(x_{k-1}+x_k)|}{k} \\
&\qquad = \frac{|x_k|}{k} \\
&\Longrightarrow x_1^2 + (x_1+x_2)^2 + \ldots + (x_{k-1}+x_k)^2 \geq \frac{x_k^2}{k}.
\end{aligned}$$

Similarly, we have

$$(x_k + x_{k+1})^2 + \cdots + (x_{n-1} + x_n)^2 + x_n^2 \geq \frac{x_k^2}{n-k+1}.$$

Adding the above two inequalities, we have

$$\begin{aligned} 2 &= x_1^2 + (x_1 + x_2)^2 + \cdots + (x_{n-1} + x_n)^2 + x_n^2 \\ &\geq \left(\frac{1}{k} + \frac{1}{n-k+1}\right) x_k^2, \end{aligned}$$

i.e., $|x_k| \leq |x_k|_{\max}$. Equality holds if and only if

$$x_1 = -(x_1 + x_2) = x_2 + x_3 = \cdots = (-1)^{k-1}(x_{k-1} + x_k)$$

and

$$x_k + x_{k+1} = -(x_{k+1} + x_{k+2}) = \cdots = (-1)^{n-k} x_n.$$

Thus $|x_k|$ obtains its maximum value $|x_k|_{\max}$ if and only if

$$x_i = \begin{cases} (-1)^{k-i} \dfrac{x_k i}{k} & \text{if } i = 1, 2, \ldots, k-1, \\ (-1)^{i-k} \dfrac{x_k(n+1-i)}{n-k+1} & \text{if } i = k+1, \ldots, n. \end{cases}$$

Second Solution. We consider the quadratic form

$$\sum_{i=1}^{n} x_i^2 + \sum_{i=1}^{n-1} x_i x_{i+1}.$$

For a fixed positive k, $1 \leq k \leq n$, we complete squares from both ends towards x_k. We do this by finding real numbers $a_1, a_2, \cdots, a_n$ such that

$$\begin{aligned} 1 &= \sum_{i=1}^{n} x_i^2 + \sum_{i=1}^{n-1} x_i x_{i+1} \\ &= (\sqrt{a_1} x_1 + \sqrt{1-a_2} x_2)^2 + (\sqrt{a_2} x_2 + \sqrt{1-a_3} x_3)^2 \\ &\quad + \cdots + (\sqrt{a_{k-1}} x_{k-1} + \sqrt{1-a_k} x_k)^2 \\ &\quad + (\sqrt{1-a_{n+1-k}} x_k + \sqrt{a_{n-k}} x_{k+1})^2 \\ &\quad + \cdots + (\sqrt{1-a_2} x_{n-1} + \sqrt{a_1} x_n)^2 \\ &\quad + [1 - (1-a_k) - (1-a_{n+1-k})] x_k^2. \end{aligned}$$

Here $a_1 = 1$,

$$2\sqrt{a_i(1-a_{i+1})} = 1,$$

for $i = 1, 2, \ldots, \max(k-1, n-k)$.

From induction, we can prove that

$$a_i = \frac{i+1}{2i} \text{ and } 1 - a_i = \frac{i-1}{2i}$$

for all the i. Thus

$$[1 - (1 - a_k) - (1 - a_{n+1-k})]x_k^2 \leq 1$$

and

$$|x_k| = \sqrt{\frac{1}{1 - (1 - a_k) - (1 - a_{n+1-k})}} = |x_k|_{\max}$$

if and only if

$$\begin{aligned}\sqrt{a_1}x_1 + \sqrt{1-a_2}x_2 &= \sqrt{a_2}x_2 + \sqrt{1-a_3}x_3 = \cdots \\ &= \sqrt{a_{k-1}}x_{k-1} + \sqrt{1-a_k}x_k \\ &= \sqrt{1-a_{n+1-k}}x_k + \sqrt{a_{n-k}}x_{k+1} = \cdots \\ &= \sqrt{1-a_2}x_{n-1} + \sqrt{a_1}x_n = 0.\end{aligned}$$

Solving the above equations for $x_1, x_2, \ldots, x_{k-1}, x_{k+1}, \ldots, x_n$ yields the values of the x_i's which produce the desired maximum value $|x_k|_{\max}$ for $|x_k|$.

1.4 Czech and Slovak Republics

Problem 1

Find all real numbers x such that

$$x\lfloor x\lfloor x\lfloor x\rfloor\rfloor\rfloor = 88.$$

Solution. Let $f(x) = x\lfloor x\lfloor x\lfloor x\rfloor\rfloor\rfloor$.

Lemma 8. *Let a and b be real numbers. If a and b have the same sign and $|a| > |b| \geq 1$, then $|f(a)| > |f(b)|$.*

Proof. We notice that $|\lfloor a\rfloor| \geq |\lfloor b\rfloor| \geq 1$. Multiplying this by $|a| > |b| \geq 1$, we have $|a\lfloor a\rfloor| > |b\lfloor b\rfloor| \geq 1$. Notice that $a\lfloor a\rfloor$ and $a\lfloor a\lfloor a\rfloor\rfloor$ have the same signs as $b\lfloor b\rfloor$ and $b\lfloor b\lfloor b\rfloor\rfloor$ respectively. In a similar manner,

$$|a\lfloor a\lfloor a\rfloor\rfloor| > |b\lfloor b\lfloor b\rfloor\rfloor| \geq 1, \quad |\lfloor a\lfloor a\lfloor a\rfloor\rfloor\rfloor| \geq |\lfloor b\lfloor b\lfloor b\rfloor\rfloor\rfloor| \geq 1,$$

and $|f(a)| > |f(b)|$, as claimed. □

We have $f(x) = 0$ for $|x| < 1$, $f(1) = f(-1) = 1$. Suppose that $f(x) = 88$. So $|x| > 1$, and we consider the following two cases.

(a) $x \geq 1$. It is easy to check that $f(22/7) = 88$. From the lemma, we know that $f(x)$ is increasing for $x > 1$. So $x = 22/7$ is the unique solution on this interval.

(b) $x \leq -1$. From the lemma, $f(x)$ is decreasing for $x < 1$. Since

$$|f(-3)| = 81 < f(x) = 88 < |f(-112/37)| = 112,$$

$-3 > x > -112/37$ and $\lfloor x\lfloor x\lfloor x\rfloor\rfloor\rfloor = -37$. But then $x = -88/37 > -3$, a contradiction. So there is no solution on this interval.

Therefore, $x = 22/7$ is the only solution.

Note. $22/7$ and $-112/37$ are found by finding $\lfloor x\rfloor$, $\lfloor x\lfloor x\rfloor\rfloor$, and $\lfloor x\lfloor x\lfloor x\rfloor\rfloor\rfloor$ in that order. For example, for $x \geq 1$, $f(3) < 88 < f(4)$ so $3 < x < 4$. Then $\lfloor x\rfloor = 3$ and $x\lfloor x\lfloor 3x\rfloor\rfloor = 88$. Then $f(3) < 88 < f(10/3)$ so $\lfloor x\lfloor x\rfloor\rfloor = 9$, and so on.

Problem 2

Show that from any fourteen distinct natural numbers, there exists $k \in \{1, \ldots, 7\}$ for which one can find disjoint k-element subsets $\{a_1, \ldots, a_k\}$

and $\{b_1, \ldots, b_k\}$ of the fourteen numbers such that the sums

$$A = \frac{1}{a_1} + \cdots + \frac{1}{a_k}, \quad B = \frac{1}{b_1} + \cdots + \frac{1}{b_k}$$

differ by less than 0.001.

Solution. Consider the $\binom{14}{7} = 3432$ 7-element subsets of the fourteen numbers, and look at the sums of the reciprocals of the numbers in each subset. Each sum is at most

$$1 + 1/2 + \cdots + 1/7 = 363/140 < 2.60,$$

so each of the 3432 sums lies in one of the 2600 intervals

$$(0, 1/1000],\ (1/1000, 2/1000],\ \ldots,\ (2599/1000, 2600/1000].$$

Then by the Pigeonhole Principle, some two sums lie in the same interval. Taking the corresponding sets and discarding any possible common elements, we obtain two satisfactory subsets A and B.

Problem 3

A sphere is inscribed in a given tetrahedron $ABCD$. The four planes tangent to the sphere parallel to (but distinct from) the planes of the four faces of $ABCD$ cut off four smaller tetrahedra. Prove that the sum of the lengths of the edges of the four smaller tetrahedra is twice the sum of the lengths of the edges of $ABCD$.

Solution. Let V be the volume of $ABCD$; let $S_1 = [BCD]$, $S_2 = [CDA]$, $S_3 = [DAB]$, $S_4 = [ABC]$; let h_1, h_2, h_3, h_4 be the heights from A, B, C, D; and let r be the inradius of the tetrahedron. Then

$$V = \frac{r(S_1 + S_2 + S_3 + S_4)}{3},$$

and for $i = 1, 2, 3, 4$, $V = S_i h_i/3$. So

$$\frac{r}{h_i} = \frac{S_i}{S_1 + S_2 + S_3 + S_4}$$

$$\frac{r}{h_1} + \frac{r}{h_2} + \frac{r}{h_3} + \frac{r}{h_4} = 1.$$

Now, let p be the perimeter of $ABCD$. Each smaller tetrahedron is homothetic to $ABCD$ and has height $h_i - 2r$, so its perimeter is

$p(h_i - 2r)/h_i$. Then the sum of the four perimeters is

$$4p - 2p\left(\frac{r}{h_1} + \frac{r}{h_2} + \frac{r}{h_3} + \frac{r}{h_4}\right) = 4p - 2p = 2p,$$

as desired.

Problem 4

A point A is given outside of a circle k in the plane. Show that the diagonals of any trapezoid inscribed in k, the extensions of whose nonparallel sides meet at A, meet at a point independent of the choice of trapezoid.

First Solution. Let $STUV$ be the trapezoid with $ST > UV, ST \parallel UV$; let B and C be the midpoints of ST and UV, respectively; let the diagonals TV and SU meet at P; and let O be the center of k. Also, let ℓ be the line through P parallel to ST and UV, meeting TU and SV at E and F, respectively.

Now, TU and SV meet at A, and $STUV$ must be isosceles. From parallel lines,

$$\frac{UE}{ET} = \frac{CP}{PB}$$

and

$$\frac{UA}{AT} = \frac{CA}{AB} = \frac{UC}{TB}.$$

But from similar triangles CUP and BTP (again by parallel lines), these fractions are equal. So, $(UTEA)$ is a harmonic range. Similarly, $(VSFA)$ is a harmonic range. In other words, line EF is the pole of A under the polar map with respect to k.

Therefore, P is the intersection of this pole and AC, independent of the choice of trapezoid.

Second Solution. Define $STUV$ and all other points as in the first solution. Let $AV = v$, $AS = s$, $AO = o$, and r be the radius of k; let $VC/SB = x$; let $\angle SAC = \alpha$. From parallel lines, $v = xs$, $VS = (1-x)s$,

$$BP = \frac{BC}{1+x} = \frac{VS\cos\alpha}{1+x} = \frac{(1-x)s\cos\alpha}{1+x}.$$

So

$$AP = AB - BP = s\cos\alpha - \frac{(1-x)s\cos\alpha}{1+x} = \frac{2x}{1+x}s\cos\alpha.$$

From Power of a Point, $sv = o^2 - r^2$. Thus $x = (o^2 - r^2)/s^2$. Applying the law of cosines to triangle AOS, we have

$$\cos\alpha = \frac{s^2 + o^2 - r^2}{2so},$$

which impies that $AP = (o^2 - r^2)/o$, independent of the choice of $STUV$.

Problem 5

Let a, b, c be positive real numbers. Show that there exists a triangle with sides a, b, c if and only if there exist real numbers x, y, z such that

$$\frac{y}{z} + \frac{z}{y} = \frac{a}{x}, \quad \frac{z}{x} + \frac{x}{z} = \frac{b}{y}, \quad \frac{x}{y} + \frac{y}{x} = \frac{c}{z}.$$

Solution. Rewrite the above as

$$a = \frac{xy}{z} + \frac{zx}{y}, \quad b = \frac{yz}{x} + \frac{xy}{z}, \quad c = \frac{zx}{y} + \frac{yz}{x}.$$

If x, y, z exist, then some two of them share the same sign: say, x and y. Then $z = c/(x/y + y/x) > 0$. Thus $a+b-c = 2xy/z$, $b+c-a = 2yz/x$, and $c+a-b = 2zx/y$ are all positive, so a, b, c form a triangle.

Conversely, if there is a triangle with sides a, b, c, then let $u = b+c-a$, $v = c+a-b$, $w = a+b-c$; by the triangle inequality, these are all positive. If there did exist satisfactory x, y, z, then from above $u = 2yz/x$, $v = 2zx/y$, $w = 2xy/z$. Solving these equations gives $x = \sqrt{vw}/2$, $y = \sqrt{wu}/2$, $z = \sqrt{uv}/2$, and these values indeed satisfy the equations.

1.5 Hungary

Problem 1

We are given a 3×3 table with integer entries, not all of the same parity. We repeat the following operation: simultaneously replace each entry by the sum of its neighbors (those entries sharing an edge with the given entry). Regardless of the starting position, will we always eventually obtain a table with all entries even? With all entries odd?

Solution. With operations all taken modulo 2, we have

$$\begin{bmatrix} a & b & c \\ d & e & f \\ g & h & i \end{bmatrix} \longrightarrow \begin{bmatrix} b+d & a+e+c & b+f \\ a+e+g & b+d+h+f & c+e+i \\ d+h & g+e+i & h+f \end{bmatrix}$$

$$\longrightarrow \begin{bmatrix} c+g & b+h & a+i \\ d+f & 0 & d+f \\ a+i & b+h & c+g \end{bmatrix}$$

$$\longrightarrow \begin{bmatrix} b+d+h+f & c+g+a+i & b+d+h+f \\ c+g+a+i & 0 & c+g+a+i \\ b+d+h+f & a+c+g+i & b+d+h+f \end{bmatrix}$$

$$\longrightarrow \begin{bmatrix} 0 & 0 & 0 \\ 0 & 0 & 0 \\ 0 & 0 & 0 \end{bmatrix}$$

Thus regardless of the starting position, we always obtain a table with all entries even in at most 4 steps. On the other hand,

$$\begin{bmatrix} 0 & 1 & 0 \\ 1 & 0 & 1 \\ 0 & 1 & 0 \end{bmatrix}$$

has all entries even after one operation; further operations keep all the entries even.

Note that $b+d$, $h+f$, $b+d+h+f$ are the main diagonal elements after the first operation. Of these three elements, at least one is even. So, we never obtain a table with all entries odd.

Problem 2

The incircle of triangle ABC touches the corresponding sides at A', B', C'. Let the midpoint of the arc AB of the circumcircle not containing C

be C'', and define A'' and B'' similarly. Show that the lines $A'A''$, $B'B''$, $C'C''$ are concurrent.

Solution. From equal tangents, we have $BA' = BC'$, thus

$$\angle BA'C' = \frac{180 - \angle B}{2} = \frac{\angle A + \angle C}{2}.$$

Let $A''C''$ meets BC at D. We notice that AA'', BB'', CC'' bisect angles A, B, C respectively, which implies that

$$\angle BDC'' = (\overset{\frown}{BC}/2 + \overset{\frown}{AB}/2)/2 = (\angle A + \angle C)/2 = \angle BA'C'$$

and $A'C' \parallel A''C''$. Similarly, we have $B'C' \parallel B''C''$ and $C'A' \parallel C''A''$. So the corresponding sides of the triangles $A'B'C'$ and $A''B''C''$ are parallel, our results follows from a special case of Desargues' theorem.

Problem 3

Find all solutions in real numbers x, y, z to the system of equations

$$x + y + (z^2 - 8z + 14)\sqrt{x + y - 2} = 1$$

$$2x + 5y + \sqrt{xy + z} = 3.$$

Solution. The only answer is $(x, y, z) = (4, -1, 4)$. Let $u = x + y - 2$ and $v = z^2 - 8z + 14 = (z - 4)^2 - 2 \geq -2$. Rewriting the first equation, we have $u + 1 = -v\sqrt{u} \leq 2\sqrt{u}$, which implies that $(\sqrt{u} - 1)^2 \leq 0$. So, $u = 1$, $v = -2$, i.e., $z = 4$, $x + y = 3$. We substitute $x = 3 - y$ into the second equation, isolate the square root, and square both sides to obtain $9y^2 + 18y + 9 = y(3 - y) + 4$. Solving the quadratic equation, we have $y = -1/2, -1$. Only $y = -1$ leads to a valid solution $(4, -1, 4)$ to the original system.

Problem 4

Let ABC be a triangle and P, Q, R points on AB, BC, CA, respectively. The points A', B', C' on RP, PQ, QR, respectively, have the property that $AB \parallel A'B'$, $BC \parallel B'C'$, $CA \parallel C'A'$. Show that $AB/A'B' = [PQR]/[A'B'C']$.

Solution. Triangles ABC and $A'B'C'$ are similar. Let $k = AB/A'B' = BC/B'C' = CA/C'A'$. Thus $[ABC] = k^2[A'B'C']$. Let d be the distance between the parallel lines AB and $A'B'$. In trapezoid $AA'B'B$, triangles $AB'B$ and $AA'B'$ have the same height d corresponding to the respective bases AB and $A'B'$; triangles $PA'B'$ and $AA'B'$ have the same height d corresponding to the common base $A'B'$. We have

$$[AB'B] = k[AA'B'] = k[PA'B'], \quad [AA'B'B] = (k+1)[PA'B'].$$

Similarly, we have

$$[BB'C'C] = (k+1)[QB'C'] \text{ and } [CC'A'A] = (k+1)[RC'A'].$$

Thus

$$\begin{aligned}(k+1)(k-1)[A'B'C'] &= (k^2-1)[A'B'C'] \\ &= [ABC] - [A'B'C'] \\ &= [AA'B'B] + [BB'C'C] + [CC'A'A] \\ &= (k+1)([PA'B'] + [QB'C'] + [RC'A']) \\ &= (k+1)([PQR] - [A'B'C']).\end{aligned}$$

So $(k-1)[A'B'C'] = [PQR] - [A'B'C']$ and

$$\frac{[PQR]}{[A'B'C']} = k = \frac{AB}{A'B'},$$

as desired.

Problem 5

(a) For which positive integers n do there exist positive integers x, y such that

$$\operatorname{lcm}(x,y) = n!, \qquad \gcd(x,y) = 1998?$$

(b) For which n is the number of such pairs x, y with $x \le y$ less than 1998?

Solution. **(a)** Let $x = 1998a$, $y = 1998b$. So a, b are positive integers such that $a < b$, $\gcd(a,b) = 1$. We have $\operatorname{lcm}(x,y) = 1998ab = 2 \cdot 3^3 \cdot 37ab = n!$. Thus $n \ge 37$ and it is easy to see that this condition is also sufficient.

(b) The answers are $n = 37, 38, 39, 40$. We only need to consider positive integers $n \ge 37$. For $37 \le n < 41$, let $k = ab = n!/1998$. Since $\gcd(a,b) = 1$, any prime factor of k that occurs in a cannot occur in b, and vice-versa. There are 11 prime factors of k, namely 2, 3, 5, 7, 11, 13, 17, 19, 23, 29, 31. For each of those prime factors, one must decide only whether it occurs in a or in b. These 11 decisions can be made in a total of $2^{11} = 2048$ ways. However, only half of these ways will satisfy the condition $a < b$. Thus there will be a total of 1024 such pairs of (x, y) for $n = 37, 38, 39, 40$. Since 41 is a prime, we can see by a similar argument that there will be at least 2048 such pairs of (x, y) for $n \geq 41$.

Problem 6

Let x, y, z be integers with $z > 1$. Show that

$$(x+1)^2 + (x+2)^2 + \cdots + (x+99)^2 \neq y^z.$$

Solution. We prove the statement by contradiction. Suppose, on the contrary, that there are integers x, y, z such that $z > 1$, and

$$(x+1)^2 + (x+2)^2 + \cdots + (x+99)^2 = y^z.$$

We notice that

$$\begin{aligned} y^z &= (x+1)^2 + (x+2)^2 + \cdots + (x+99)^2 \\ &= 99x^2 + 2(1+2+\cdots+99)x + (1^2+2^2+\cdots+99^2) \\ &= 99x^2 + \frac{2 \cdot 99 \cdot 100}{2}x + \frac{99 \cdot 100 \cdot 199}{6} \\ &= 33(3x^2 + 300x + 50 \cdot 199), \end{aligned}$$

which implies that $3|y$. Since $z \geq 2$, $3^2|y^z$, but 3^2 does not divide $33(3x^2 + 300x + 50 \cdot 199)$, contradiction. So our assumption in fact must be false and the original statement in the problem is correct.

Problem 7

Let ABC be an acute triangle and P a point on the side AB. Let B' be a point on the ray AC and A' a point on the ray BC such that $\angle B'PA = \angle A'PB = \angle ACB$. Let the circumcircles of triangles APB' and BPA' meet at P and M. Find the locus of M as P varies along the segment AB.

Solution. Let O be the point on the perpendicular bisector of AB such that $\angle AOB = 2\angle C$ and O and C are on the different sides of AB. Let ω be the circle centered at O with radius OA. The locus of M is the minor arc AB on ω.

We direct angles modulo $180°$. Since $AB'MP$ and $PMBA'$ are cyclic, we have

$$\angle B'MP = 180° - \angle BAC$$

$$\angle PMA' = 180^\circ - \angle CBA$$
$$\angle B'MA = \angle B'PA$$
$$\angle BMA' = \angle BPA'.$$

Add the first two equations and subtract the last equations, we have

$$\angle AMB = 360^\circ - \angle BAC - \angle CBA - 2\angle ACB$$
$$= 180^\circ - \angle ACB,$$

which implies that M is the minor arc AB on ω, as claimed.

Problem 8

Let $ABCDEF$ be a centrally symmetric hexagon and P, Q, R points on sides AB, CD, EF, respectively. Show that the area of triangle PQR is at most half the area of $ABCDEF$.

First Solution. We establish the following lemma.

Lemma 9. *If $ABCDEF$ is a centrally symmetric hexagon, then*

(a) each of AD, BE, CF cuts the area of $ABCDEF$ in half;

(b) $[ACE] = [BDF] = \dfrac{[ABCDEF]}{2}$.

Proof. Part (a) follows easily from the symmetric property. Now we prove part (b). WLOG, we only prove that $[ACE] = [ABCDEF]/2$. Let O be the symmetric center. Then AD, BE, CF are concurrent at O. From the symmetry, $BO = OE$ and $[COE] = [COB]$. Similarly, we have $[OEA] = [OBA], [AOC] = [AOF]$. Add the above together, we have $[ACE] = [ABCF]$. But from (a), we know that $[ABCF] = [ABCDEF]/2$. Thus $[ACE] = [ABCDEF]/2$, as desired. □

Now we prove our main result. Since Q is on CD,

$$[PQR] \le \max\{[PCR], [PDR]\}.$$

Let $Y \in \{C, D\}$ such that $[PYR] = \max\{[PCQ], [PDQ]\}$. Then moving PQR to PYR increases the area. Repeating this for each vertex gives a triangle XYZ with at least the same area and $\{X, Y, Z\} \subset \{A, B, C, D, E, F\}$. We consider the following two cases.

(1) Some of X, Y, Z are next to each other in the original hexagon. WLOG, say $X = B$, $Y = C$. Then XYZ must be on the same side of either BE or CF. Thus from part (a) of the lemma,

$$[PQR] \leq [XYZ] < \frac{[ABCDEF]}{2},$$

as desired.

(2) If none of X, Y, Z are next to each other in the original hexagon, then XYZ is either ACE or BDF. Thus from part (b) of the lemma,

$$[PQR] \leq [XYZ] < \frac{[ABCDEF]}{2},$$

as desired.

Second Solution. Reflect PQR with respect to O, the center of the hexagon, to form triangle STU. Then S, T, U are on sides DE, FA, BC. So, $PUQSRT$ is completely enclosed by $ABCDEF$ and $[PUQSRT] \leq [ABCDEF]$. Then from part (b) of the lemma in the first solution,

$$[PQR] = \frac{[PUQSRT]}{2} \leq \frac{[ABCDEF]}{2},$$

as desired.

Problem 9

Two players take turns drawing a card at random from a deck of four cards labeled 1, 2, 3, 4. The game stops as soon as the sum of the numbers that have appeared since the start of the game is divisible by 3, and the player who drew the last card is the winner. What is the probability that the player who goes first wins?

First Solution. The answer is $13/23$. Let P be the desired probability. For positive integers n, let b_n be the probability that the n-th card has been drawn and the sum of the numbers on the first n cards is 1 modulo 3; let c_n be the probability that the n-th card has been drawn and the sum of the numbers on the first n cards is 2 modulo 3; and let a_n be the probability that the game ends immediately after the n-th card has been drawn. We notice that $a_1 = 1/4$, $b_1 = 1/2$, $c_1 = 1/4$, $b_2 = 3/16$, and we have the following relations:

$$b_{n+1} = \frac{b_n}{4} + \frac{c_n}{4}, \quad c_{n+1} = \frac{b_n}{2} + \frac{c_n}{4}, \quad a_{n+1} = \frac{b_n}{4} + \frac{c_n}{2}$$

Subtracting the first two equations from each other, we obtain that $c_{n+1} = b_n/4 + b_{n+1}$. Substituting this back to the first equation, we have $16b_{n+2} =$

$8b_{n+1} + b_n$. Solving the characteristic equation $16x^2 = 6x + 1$ we have

$$b_n = \left(\frac{3\sqrt{2}-2}{4}\right)\left(\frac{\sqrt{2}+1}{4}\right)^n - \left(\frac{3\sqrt{2}+2}{4}\right)\left(\frac{1-\sqrt{2}}{4}\right)^n.$$

We have $a_{n+2} = b_{n+1}/4 + c_{n+1}/2 = b_{n+1}/4 + (b_n/4 + b_{n+1})/2 = 3b_n/4 + b_n/8$. To solve our problem, we only need to calculate four convergent infinite series. In fact,

$$\begin{aligned} P &= a_1 + a_3 + \cdots + a_{2n-1} + \cdots \\ &= \frac{1}{4} + \frac{3(b_2 + b_4 + \cdots)}{4} + \frac{b_1 + b_3 + \cdots}{4} \\ &= \frac{1}{4} + \frac{21}{92} + \frac{2}{23} \\ &= \frac{13}{23}. \end{aligned}$$

Second Solution. Let a_1, a_2 be the probabilities that player one will win given that when he starts a turn, the sum is 1 or 2 (mod 3), respectively (in other words, as if the sum were 1 or 2 at the beginning of the game.). Let b_1 and b_2 be the probabilities that player one will win given that when his opponent starts a turn, the sum is 1 or 2 (mod 3), respectively. We wish to find

$$P = \frac{1}{4} + \frac{1}{2}b_1 + \frac{1}{4}b_2.$$

Now, we have the following relations:

$$a_1 = \frac{1 + b_1 + 2b_2}{4}, \quad a_2 = \frac{2 + b_1 + b_2}{4},$$

$$b_1 = \frac{a_1 + 2a_2}{4}, \quad b_2 = \frac{a_1 + a_2}{4}.$$

We multiply the first two equations by 4 and the last two equations by 16, then substitute the first two equations into the last two equations to get two equations in b_1 and b_2. These equations have solution $b_1 = 11/23$, $b_2 = 7/23$, which implies that $P = 13/23$.

Problem 10

Let ABC be a triangle and P, Q points on the side AB such that the inradii of triangles APC and QBC are the same. Prove that the inradii of triangles AQC and PBC are also the same.

First Solution. We first establish the following lemma.

Lemma 10. *Let XYZ be a triangle, let r be the inradius of XYZ, let h_x be the length of the altitude corresponding to base YZ. Then*

$$\frac{h_x - 2r}{h_x} = \tan\frac{\angle Y}{2}\tan\frac{\angle Z}{2}.$$

Proof. Let ω and I be the incircle and the incenter of XYZ, respectively. Let ω and YZ meet at S. Let Y_1 and Z_1 be points on XY and XZ, respectively, such that Y_1Z_1 is tangent to ω at T and $Y_1Z_1 \parallel YZ$. We have

$$\angle TY_1I = \frac{180^\circ - \angle XYZ}{2}, \quad \angle TZ_1I = \frac{180^\circ - \angle XZY}{2},$$

$$\Longrightarrow \quad \frac{h_x - 2r}{h_x} = \frac{Y_1Z_1}{YZ} = \frac{\cot\angle TY_1I + \cot\angle TZ_1I}{\cot\frac{\angle XYZ}{2} + \cot\frac{\angle XZY}{2}}$$

$$= \frac{\tan\frac{\angle XYZ}{2} + \tan\frac{\angle XZY}{2}}{\cot\frac{\angle XYZ}{2} + \cot\frac{\angle XZY}{2}}$$

$$= \tan\frac{\angle XYZ}{2}\tan\frac{\angle XZY}{2},$$

as desired. □

Now we prove our main result. Let h_c be the length of the altitude corresponding to side AB. Let r_1, r_2, r_3, r_4 be the inradii of triangles CAP, CBQ, CAQ, CBP, respectively. Since $r_1 = r_2$, from the lemma, we have

$$\frac{h_c - 2r_1}{h_c} = \frac{h_c - 2r_2}{h_c}$$

$$\tan\frac{\angle CAP}{2}\tan\frac{\angle APC}{2} = \tan\frac{\angle CBQ}{2}\tan\frac{\angle BQC}{2}$$

$$\tan\frac{\angle CAP}{2}\cot\frac{\angle BPC}{2} = \tan\frac{\angle CBQ}{2}\cot\frac{\angle AQC}{2}$$

$$\tan\frac{\angle CAQ}{2}\tan\frac{\angle AQC}{2} = \tan\frac{\angle CBP}{2}\tan\frac{\angle BPC}{2}$$

$$\frac{h_c - 2r_3}{h_c} = \frac{h_c - 2r_4}{h_c}$$

$$r_3 = r_4,$$

as desired.

Second Solution. We provide an alternate proof to the lemma. Define ω, I, Y_1, Z_1 as before. Triangles XY_1Z_1 and XYZ are similar, so the ratio $(h_x - 2r)/(h_x)$ is the ratio between the corresponding altitudes of the triangles, and hence it is equal the ratio between the inradii of the triangles.

Let r_1, K_1, and s_1 be the inradius, area, and semiperimeter of triangle XY_1Z_1, respectively. Also let $x_1 = Y_1Z_1$, $y_1 = Z_1X$, and $z_1 = Y_1X$. Since ω is an excircle of triangle XY_1Z_1, we have

$$K_1 = r(s_1 - x_1) = r_1 s_1 = \sqrt{s_1(s_1 - x_1)(s_1 - y_1)(s_1 - z_1)}.$$

Together, these equations imply that

$$\frac{r_1}{r} = \frac{s_1 - x_1}{s_1} = \frac{r_1}{s_1 - y_1} \cdot \frac{r_1}{s_1 - z_1} = \tan\frac{\angle Y}{2}\tan\frac{\angle Z}{2},$$

as desired.

1.6 India

Problem 1

(a) Show that the product of two numbers of the form $a^2 + 3b^2$ is again of that form.

(b) If an integer n is such that $7n$ is of the form $a^2 + 3b^2$, prove that n is also of that form.

Solution. **(a)** $(a^2 + 3b^2)(c^2 + 3d^2) = (ac + 3bd)^2 + 3(ad - bc)^2$.

(b) $7 = (\pm 2)^2 + 3(1)^2$, so reversing the above construction may yield the desired result. Indeed, if $7n = e^2 + 3f^2$, then solving for c and d after setting $a = 2, b = 1$ and $a = -2, b = 1$ gives

$$c_1 = \frac{2e - 3f}{7},\ d_1 = \frac{e + 2f}{7};\ c_2 = \frac{2e + 3f}{7},\ d_2 = \frac{e - 2f}{7}.$$

With all congruences taken modulo 7, we have

$$e^2 + 3f^2 \equiv 0 \iff e^2 \equiv 4f^2 \iff e \equiv \pm 2f.$$

If $e \equiv -2f$, then c_1 and d_1 are integers; if $e \equiv 2f$, then c_2 and d_2 are integers. Either way, n can be written in the desired form.

Problem 2

Let a, b, c be three real numbers and let

$$X = a + b + c + 2\sqrt{a^2 + b^2 + c^2 - ab - bc - ca}.$$

Show that $X \geq \max\{3a, 3b, 3c\}$, and that one of the numbers

$$\sqrt{X - 3a},\ \sqrt{X - 3b},\ \sqrt{X - 3c}$$

is the sum of the other two.

Solution. We have $3(b - c)^2 \geq 0$, which implies that $4b^2 + 4c^2 - 4bc \geq b^2 + c^2 + 2bc$. Thus

$$\begin{aligned} 4(a^2 + b^2 + c^2 - ab - bc - ca) &\geq (2a - b - c)^2 \\ &\implies 2\sqrt{a^2 + b^2 + c^2 - ab - bc - ca} \\ &\geq |2a - b - c| \geq 2a - b - c \\ &\implies X \geq 3a, \end{aligned}$$

Similarly, $X \geq 3b$ and $X \geq 3c$, so $X \geq \max\{3a, 3b, 3c\}$. Now, let $u = \sqrt{X - 3a}$, $v = \sqrt{X - 3b}$, $w = \sqrt{X - 3c}$. Then $a = (X - u^2)/3$, $b = (X - v^2)/3$, and $c = (X - w^2)/3$. We have

$$
\begin{aligned}
& X = a + b + c + 2\sqrt{a^2 + b^2 + c^2 - ab - bc - ca} \\
\Longleftrightarrow\ & X = a + b + c + 2\sqrt{\frac{1}{2}(a-b)^2 + \frac{1}{2}(b-c)^2 + \frac{1}{2}(c-a)^2} \\
\Longleftrightarrow\ & u^2 + v^2 + w^2 = 2\sqrt{\frac{1}{2}(u^2 - v^2)^2 + \frac{1}{2}(v^2 - w^2)^2 + \frac{1}{2}(w^2 - u^2)^2} \\
\Longleftrightarrow\ & u^2 + v^2 + w^2 = 2\sqrt{u^4 + v^4 + w^4 - u^2v^2 - v^2w^2 - w^2u^2} \\
\Longleftrightarrow\ & u^4 + v^4 + w^4 - 2u^2v^2 - 2v^2w^2 - 2w^2u^2 = 0 \\
\Longleftrightarrow\ & u^4 + 2u^2v^2 + v^4 - 2w^2(u^2 + v^2) + w^4 - 4u^2v^2 = 0 \\
\Longleftrightarrow\ & (u^2 + v^2)^2 - 2w^2(u^2 + v^2) + w^4 - 4u^2v^2 = 0 \\
\Longleftrightarrow\ & (u^2 + v^2 - w^2)^2 - (2uv)^2 = 0 \\
\Longleftrightarrow\ & (u^2 + v^2 + 2uv - w^2)(u^2 + v^2 - 2uv - w^2) = 0 \\
\Longleftrightarrow\ & ((u+v)^2 - w^2)(u-v)^2 - w^2) = 0 \\
\Longleftrightarrow\ & (u+v+w)(u+v-w)(u-v-w)(u-v-w) = 0.
\end{aligned}
$$

From definition, $u, v, w \geq 0$. If $uvw = 0$, then $X = 3a = 3b = 3c$ and we are done. If $uvw \neq 0$, then $u + v + w > 0$ and one is the sum of the other two, as desired.

Problem 3

Let n be a positive integer, and let X be a set of $n + 2$ integers each of absolute value at most n. Show that there exist three distinct numbers a, b, c in X such that $a + b = c$.

Solution. We prove the claim by induction.

First, $n = 1$, our set must be $\{-1, 0, 1\}$ and $-1 + 1 = 0$.

Now suppose that the claim is true for $n = k - 1$. For $n = k$, we prove our claim by contradiction. Suppose, on the contrary, that there is a set X disproving the claim for $n = k$. If k is in X, then at most one number from each of the pairs $\{-k, 0\}, \{-k+1, 1\}, \ldots, \{-1, k-1\}$ can be in X. So there are at most $k + 1$ integers in X, a contradiction. Thus, k is not in X; similarly, nor is $-k$.

But then any $k+1$ integers of X form a set X_1 of integers satisfying the conditions for $n = k - 1$. By the induction hypothesis, there are three distinct numbers a, b, c in X_1 (and thus X) such that $a + b = c$, a contradiction.

Therefore, the claim is true for $n = k$; and by induction, the claim is true for all positive integers n.

Problem 4

In triangle ABC, let AK, BL, CM be the altitudes and H the orthocenter. Let P be the midpoint of AH. If BH and MK meet at S, and LP and AM meet at T, show that TS is perpendicular to BC.

Solution. All angles are directed modulo 180°. Since $\angle BMH = 90° = \angle BKH$, $BMHK$ is cyclic and thus

$$\angle LHP = \angle BHK = \angle BMK$$

Since LP is the median to the hypotenuse of right triangle ALH, we have $LP = HP$ and thus

$$\angle TLS = \angle PLH = \angle LHP.$$

Finally, we have

$$\angle BMK = \angle BMS = 180° - \angle SMT.$$

Together, these equations imply that $\angle TLS = 180° - \angle SMT$ so that $TLSM$ is cyclic. Then $\angle TSL = \angle TML = \angle AML$, which equals $\angle AHL$ since $AMHL$ is cyclic as $\angle AMH = 90° = \angle ALH$. Therefore, TS is parallel to AH and is perpendicular to BC, as desired.

Problem 5

Find the number of integers x with $|x| \leq 1997$ such that 1997 divides $x^2 + (x+1)^2$.

Solution. There are 4 such integers. With congruences all taken modulo 1997, we have

$$x^2 + (x+1)^2 \equiv 2x^2 + 2x + 1 \equiv 4x^2 + 4x + 2 \equiv 0,$$

i.e., $(2x+1)^2 \equiv -1$. Since 1997 is a prime of the form $4k+1$, there are exactly two distinct solutions to $u^2 \equiv -1$. Each corresponds to a different solution to $(2x+1)^2 \equiv -1$.

Also, the two solutions to $(2x+1)^2 \equiv -1$ are nonzero since 0 does not satisfy the equation. Therefore, there are exactly two satisfactory integers x from -1997 to -1 and two more from 1 to 1997, for a total of four integer solutions, as claimed.

Problem 6

Let $ABCD$ be a parallelogram. A circle lying inside $ABCD$ touches the lines AB and AD, and intersects BD at E and F. Show that there exists a circle passing through E and F and touching the lines CB and CD.

Solution. Let k_1 be the given circle, P be its center, and T_1 and T_2 be the feet of the perpendiculars from P to AB and AD respectively, i.e., T_1 and T_2 are the points of tangency. Let $a = AB = CD$, $b = BC = DA$. By equal tangents, let $x = AT_1 = AT_2$.

Now, let T_3 be on ray CB so that $BT_3 = BT_1$ but T_3 is not on segment CB. Draw the perpendicular to CB through T_3 and let it intersect the the internal angle bisector of BCD at Q. We claim that the circle k_2 centered at Q with radius QT_3 is a satisfactory circle.

Since Q is on the internal angle bisector of BCD and QT_3 is perpendicular to CB, k_2 tangent to both CB and CD, so we need only show it intersects BD at E and F. Let T_4 be the point of tangency between k_2 and CD. We have $BT_3 = BT_1 = AB - AT_1 = a - x$. By equal tangents, $CT_4 = CT_3 = CB + BT_3 = b + a - x$. Since $a > x$, we have $CT_4 > CD$ and D is in between C and T_4. Then

$$DT_4 = CT_4 - CD = (b + a - x) - a = b - x = DT_2.$$

Thus, we have $BT_1 = BT_3$ and $DT_2 = DT_4$. Since BT_1 and BT_3 are tangents from B to k_1 and k_2, B has equal power with respect to the two circles. Similarly, so does D, so BD is the radical axis of k_1 and k_2. Therefore, since k_1 intersects BD at E and F, so does k_2, as desired.

Problem 7

Find all triples (x, y, n) of positive integers such that

$$\gcd(x, n+1) = 1 \quad \text{and} \quad x^n + 1 = y^{n+1}.$$

Solution. All solutions are of the form $(a^2 - 1, a, 1)$ with a even. We have $x^n = y^{n+1} - 1 = (y-1)m$ with $m = y^n + y^{n-1} + \cdots + y + 1$.

Thus $m|x^n$ and $\gcd(m, n+1) = 1$. Rewrite m as

$$m = (y-1)(y^{n-1} + 2y^{n-2} + 3y^{n-3} + \cdots + (n-1)y + n) + (n+1)$$

we have $(n+1)|\gcd(m, y-1)$. But $\gcd(m, n+1) = 1$, so $\gcd(m, y-1) = 1$. Since $x^n = (y-1)m$, m must be a perfect nth power. But

$$(y+1)^n = y^n + \binom{n}{1}y^{n-1} + \cdots + \binom{n}{n-1}y + 1 > f > y^n,$$

for $n > 1$. So m can be a perfect nth power only if $n = 1$ and $x = y^2 - 1$. Since x and $n + 1 = 2$ are relatively prime, y must be even, yielding the presented solutions.

Problem 8

Let M be a positive integer and consider the set

$$S = \{n \in \mathbb{N} : M^2 \leq n < (M+1)^2\}.$$

Prove that the products of the form ab with $a, b \in S$ are all distinct.

Solution. We prove the statement by contradiction. Suppose, on the contrary, that $\{a, b\}$ and $\{c, d\}$ are distinct subsets of S such that $ab = cd$. Assume without loss of generality that $a < c, d$.

Let $p = \gcd(a, c)$, $q = a/p$, and $r = c/p$; then $\gcd(q, r) = 1$. Since $q|(ab/p) = cd/p = rd$, $q|d$. Now let $s = d/q$ so that $b = cd/a = rs$. Therefore, $a = pq$, $b = rs$, $c = pr$, $d = qs$ for some positive integers p, q, r, s.

Since $c > a$, $r > q$ and $r \geq q + 1$. Since $d > a$, $s > p$ and $s \geq p + 1$. Therefore,

$$\begin{aligned} b = rs &\geq (p+1)(q+1) &= pq + p + q + 1 \\ &\geq pq + 2\sqrt{pq} + 1 &= a + 2\sqrt{a} + 1 \\ &\geq M^2 + 2M + 1 &= (M+1)^2. \end{aligned}$$

Then b is not in S, a contradiction. Therefore, the products of the described form are distinct, which was to be proved.

Note. If S also contained $(M+1)^2$, the claim would not be true; if $a = M^2$, $b = (M+1)^2$, and $c = d = M(M+1)$, then $ab = cd$. Also, this question is a special case of St. Petersburg 1998/22 with $n = 2M$.

Problem 9

Let N be a positive integer such that $N+1$ is prime. Choose $a_i \in \{0,1\}$ for $i=0,\dots,N$. Suppose that the a_i are not all equal, and let $f(x)$ be a polynomial such that $f(i)=a_i$ for $i=0,\dots,N$. Prove that the degree of $f(x)$ is at least N.

Solution. With all congruences taken modulo prime $p=N+1$. We notice that $\binom{p-1}{i}$ is an integer and

$$\binom{p-1}{i} = \frac{(p-1)(p-2)\cdots(p-i)}{1\cdot 2\cdots i}$$
$$\equiv \frac{(-1)(-2)\cdots(-i)}{1\cdot 2\cdots i} \equiv (-1)^i.$$

We prove the statement by contradiction. Suppose, on the contrary, that $f(x)$ has degree less than N. Any polynomial $g(x)$ of degree d or less satisfies the recursion

$$\sum_{i=0}^{d+1}(-1)^i\binom{d+1}{i}g(x+i)=0.$$

$f(x)$ has degree $N-1$ or less; applying this recursion with $g(x)=f(x)$, $x=0$, $d+1=N=p-1$, we have

$$0=\sum_{i=0}^{p-1}(-1)^i\binom{p-1}{i}f(i)\equiv\sum_{i=0}^{p-1}(-1)^i(-1)^i f(i)\equiv\sum_{i=0}^{p-1}f(i).$$

But since the $f(i)$ are all 0 or 1 but not all equal, this is impossible, a contradiction. Therefore, the degree of $f(x)$ is at least N, which was to be proved.

Problem 10

Let n, p be positive integers with $3\le p\le n/2$. Consider a regular n-gon with p vertices colored red and the rest colored blue. Show that there are two congruent nondegenerate polygons each with at least $\lfloor p/2\rfloor+1$ vertices, such that the vertices of one polygon are colored red and the vertices of the other polygon are colored blue.

Solution. We prove the statement by contradiction. Suppose, on the contrary, that there are no such two polygons. Consider the n polygons formed by rotations of the red p-gon. In the original p-gon, there are

p red vertices. In the other $n-1$ polygons, there must be at least $p - \lfloor p/2 \rfloor = \lceil p/2 \rceil$ red vertices or else the blue vertices in this p-gon would form a $(\lfloor p/2 \rfloor + 1)$-gon congruent to a red polygon. Thus, we count at least

$$(n-1)\lceil p/2 \rceil + p > n\lceil p/2 \rceil$$

red vertices. Each red vertex is counted p times (one for each position of the p-gon), so we must have more than

$$n\lceil p/2 \rceil / p \geq n/2$$

red vertices, a contradiction. Therefore, there are two congruent polygons - one blue, one red - which was to be proved.

Problem 11

Let P be a point in the interior of the convex quadrilateral $ABCD$. Show that at least one of the angles $\angle PAB, \angle PBC, \angle PCD, \angle PDA$ is less than or equal to $\pi/4$.

First Solution. This is a generalization of a special property of the Brocard points of a triangle. The following solution is similar to those of IMO 1991/5 and AIME 1999/14 in their respective pamphlets. Let $a = AB$, $b = BC$, $c = CD$, $d = DA$; $w = AP$, $x = BP$, $y = CP$, $z = DP$; $\alpha = \angle PAB$, $\beta = \angle PBC$, $\gamma = \angle PCD$, $\delta = \angle PDA$.

Let XYZ be a triangle and $XY = z$, $YZ = x$, $ZX = y$. By "smoothing" pairs of adjacent side lengths of XYZ to their average, we keep the perimeter constant while not decreasing the triangle's area. Actually we can fix X and Y and find a point Z' such that Z' is on the perpedicular bisector of XY and $2Z'X = x + y$. Applying Heron's formula, we have

$$16[XYZ]^2 = (y+z-x)(z+x-y)(x+y-z)(x+y+z)$$
$$16[XYZ']^2 = z^2(x+y-z)(x+y+z).$$

Note that

$$(y+z-x)(z+x-y) = z^2 - (x-y)^2 \leq z^2$$

gives $[XYZ] \leq [XYZ']$.

So by smoothing AB and BC; CD and DA; BC and CD; and DA and AB in that order, we form a rhombus then a square with the same perimeter and at least as much as area as the original quadrilateral. The area of this square is

$$\left(\frac{a+b+c+d}{4}\right)^2 \leq \frac{a^2+b^2+c^2+d^2}{4}$$

by the root-mean-square inequality. Thus

$$\begin{aligned} a^2+b^2+c^2+d^2 &\geq 4[ABCD] \\ &= 4([PAB]+[PBC]+[PCD]+[PDA]) \\ &= 2(aw\sin\alpha + bx\sin\beta + cy\sin\gamma + dz\sin\delta). \end{aligned}$$

Also, applying the law of cosines on triangles PAB, PBC, PCD, PDA on angles α, β, γ, δ, we have

$$x^2 = a^2 + w^2 - 2aw\cos\alpha$$

and three other similar expressions. Summing these gives

$$\begin{aligned} &2(aw\cos\alpha + bx\cos\beta + cy\cos\gamma + dz\cos\delta) \\ &\quad = a^2+b^2+c^2+d^2 \\ &\quad \geq 2(aw\sin\alpha + bx\sin\beta + cy\sin\gamma + dz\sin\delta). \end{aligned}$$

Now when $45^\circ < \theta < 180^\circ$, $\sin\theta > \cos\theta$. So if $\alpha, \beta, \gamma, \delta$ were all greater than 45° then the above inequality would be false. Therefore one must be less than or equal to 45°, as desired.

Second Solution. We are going to use a more algebraic approach to prove

$$4[ABCD] \leq a^2+b^2+c^2+d^2$$

and the rest is the same. Under the same notations, we have

$$2[ABCD] = 2[ABC] + 2[ACD] = ab\sin B + cd\sin D$$

and

$$4[ABCD]^2 = a^2b^2\sin^2 B + c^2d^2\sin^2 D + 2abcd\sin B\sin D$$

Applying law of cosines to triangles ABC and ACD, we have

$$a^2+b^2-2ab\cos B = AC^2 = c^2+d^2-2cd\cos D,$$

which implies that

$$\frac{(a^2+b^2-c^2-d^2)^2}{4} = a^2b^2\cos^2 B + c^2d^2\cos^2 D - 2abcd\cos B\cos D.$$

Thus

$$4[ABCD]^2 + \frac{(a^2+b^2-c^2-d^2)^2}{4} = a^2b^2 + c^2d^2 - 2abcd\cos(B+D)$$

$$\Longrightarrow 4[ABCD]^2 \le a^2b^2 + c^2d^2 + 2abcd = (ab+cd)^2$$

$$\Longrightarrow 2[ABCD] \le ab + cd \le \frac{a^2+b^2}{2} + \frac{c^2+d^2}{2}$$

$$\Longrightarrow 4[ABCD] \le a^2+b^2+c^2+d^2.$$

Problem 12

Let $\alpha_1, \alpha_2, \ldots, \alpha_n$ be complex numbers and put

$$f(z) = \prod_{i=1}^{n}(z - \alpha_i).$$

Show that there exists a complex z_0 with $|z_0| = 1$ such that

$$|f(z_0)| \ge \frac{\prod_{j=1}^{n}(1+|\alpha_j|)}{2^{n-1}}.$$

Solution. First suppose that the α_j lie on the unit circle; then we wish to prove there exists a z_0 with $|z_0| = 1$ such that $|f(z_0)| \ge 2$. Write $f(z) = z^n + c_{n-1}z^{n-1} + \cdots + c_1 z + c_0$, we have $c_0 = \alpha_1\alpha_2\cdots\alpha_n$ and $|c_0| = 1$. Let w be a complex number such that $w^n = c_0$. For $j = 0, \ldots, n-1$, let $w_j = e^{2ij\pi/n}$, the nth root of unity and $x_j = ww_j$. We have $|x_j| = |ww_j| = 1$ and

$$\begin{aligned}
&|f(x_0)| + \cdots + |f(x_{n-1})| \\
&\quad \ge |f(x_0) + \cdots + f(x_{n-1})| \\
&\quad = \left|\sum_{j=0}^{n-1}(x_j^n + cn - 1x_j^{n-1} + \cdots + c_1 x_j + c_0)\right| \\
&\quad = \left|w^n \sum_{j=0}^{n-1} w_j^n + c_{n-1}w^{n-1}\sum_{j=0}^{n-1} w_j^{n-1} + \cdots + c_1 w \sum_{j=0}^{n-1} w_j^1 + nc_0\right| \\
&\quad = |nw^n + 0 + \cdots + 0 + nc_0| = |2nc_0| = 2n.
\end{aligned}$$

So, one of the $|f(x_j)| \ge 2$, as desired.

Now suppose the α_j do not all lie on the circle. Let $\beta_j = \alpha_j/|\alpha_j|$. We claim that for any j and any complex number z on the unit circle,

$$\frac{|z-\alpha_j|}{1+|\alpha_j|} \geq \frac{|z-\beta_j|}{1+|\beta_j|}.$$

Let Z, A, B, B' be the points where z, α_j, β_j, and $-\beta_j$ are. If Z lies on line AB, then it equals either B or B'. In the first case, the left side is nonnegative and the right side is 0; in the second case, both sides equal one.

Otherwise, look at the triangle formed by Z, A, and B'. Applying law of sines to triangle ZAB' and right triangle ZBB', we have

$$\begin{aligned}\frac{|z-\alpha_j|}{(1+|\alpha_j|} &= \frac{ZA}{AB'} = \frac{\sin\angle ZB'A}{\sin\angle B'ZA}\\ &\geq \sin\angle ZB'B = \frac{ZB}{BB'} = \frac{|z-\beta_j|}{1+|\beta_j|}\end{aligned}$$

as claimed. Now, from the beginning of the proof, there exists a z_0 such that

$$\frac{1}{2^{n-1}} \leq \frac{\prod_{j=1}^{n}(z_0-\beta_j)}{\prod_{j=1}^{n}(1+|\beta_j|)} \leq \frac{\prod_{j=1}^{n}(z_0-\alpha_j)}{\prod_{j=1}^{n}(1+|\alpha_j|)}.$$

Problem 13

Let m, n be natural numbers with $m \geq n \geq 2$. Show that the number of polynomials of degree $2n-1$ with distinct coefficients from the set $\{1, 2, \ldots, 2m\}$ which are divisible by $x^{n-1} + \cdots + x + 1$ is

$$2^n n! \left(4\binom{m+1}{n+1} - 3\binom{m}{n}\right).$$

Solution. For $n \geq 2$, let $h(x) = x^{n-1} + \cdots + x + 1$ and let

$$f(x) = a_{2n-1}x^{2n-1} + \cdots + a_1 x + a_0$$

be a degree $(2n-1)$ polynomial. First we claim that $f(x)$ is divisible by $h(x)$ if and only if

$$a_{2n-1} + a_{n-1} = a_{2n-2} + a_{n-2} = \cdots = a_n + a_0.$$

For $1 \leq i \leq n-1$, let w_i be distinct the nth roots of unity other than 1. Since $h(x)(x-1) = x^n - 1$, the w_i are the roots to $h(x)$. So, $h(x) \mid f(x)$ if and only if $f(w_i) = 0$ for all w_i. Let

$$g(x) = (a_{2n-1} + a_{n-1})x^{n-1} + (a_{2n-2} + a_{n-2})x^{n-2} + \cdots + (a_n + a_0).$$

Since $w_i^n = 1$, we have $w_i^{2n-1} = w_i^{n-1}$, $w_i^{2n-2} = w_i^{n-2}, \cdots$, and $w_i^n = w_i^0$. Then $g(w_i) = f(w_i)$ for all i, so $h(x)$ divides $f(x)$ if and only if it divides $g(x)$. But this occurs if and only if

$$a_{2n-1} + a_{n-1} = a_{2n-2} + a_{n-2} = \cdots = a_n + a_0,$$

as claimed.

Suppose that there are N sets of n pairs of distinct integers from $1, 2, \ldots, 2m$ such that the n pairs have the same sum. In any set, the pairs are disjoint since the numbers in each pair have the same sum. Then there are N sets to choose from and $n!$ ways to choose which pairs from the set correspond to which pairs (a_i, a_{i-n}). Also, for each of the n pairs, there are 2 ways to assign the values. Therefore, there are a total of $2^n n! N$ satisfactory polynomials. Now we only need to prove that

$$N = 4\binom{m+1}{n+1} - 3\binom{m}{n}.$$

Let $S = \{1, 2, \ldots, 2m\}$. For a positive integer $k \leq 2m$, there are $\lfloor (k-1)/2 \rfloor$ pairs of distinct integers from S that add up to k: the $k-1$ pairs $\{1, k-1\}, \{2, k-2\}, \ldots, \{k-1, 1\}$ count each pair twice as well as a possible $\{k/2, k/2\}$ pair. If $4m > k > 2m$, then there are $\lfloor (4m-k+1)/2 \rfloor$ pairs of distinct integers from $[2m]$ that add up to k by a similar argument. As k ranges from $3 = 1 + 2$ to $4m - 1 = (2m-1) + 2m$, the numbers of pairs of distinct integers from $[2m]$ adding up to k are

$$1, 1, 2, 2, \ldots, m-1, m-1, m, m-1, m-1, \ldots, 2, 2, 1, 1.$$

Now, a set of n pairs of distinct integers from $1, 2, \ldots, 2m$ that share the same sum is simply a set of n pairs of the $\lfloor (k-1)/2 \rfloor$ or $\lfloor (4m-k+1)/2 \rfloor$ pairs above. So the number of such sets is equal to

$$\begin{aligned} 2&\binom{1}{n} + 2\binom{2}{n} + \cdots + \binom{m}{n} + \cdots + 2\binom{2}{n} + 2\binom{1}{n} \\ &= 4\left(\binom{1}{n} + \binom{2}{n} + \cdots + \binom{m-1}{n} + \binom{m}{n}\right) - 3\binom{m}{n} \\ &= 4\binom{m+1}{n+1} - 3\binom{m}{n}, \end{aligned}$$

and this completes our proof.

1.7 Iran

Problem 1

Let KL and KN be lines tangent to the circle C, with $L, N \in C$. Choose M on the extension of KN past N, and let P be the second intersection of C with the circumcircle of KLM. Let Q be the foot of the perpendicular from N to ML. Prove that $\angle MPQ = 2\angle KML$.

First Solution. Let S be the second intersection of C and LM; let line PS meet MN at R. From the tangencies and cyclic quadrilaterals, we have $\angle KLM = \angle KLS = \angle LPS$ and

$$\begin{aligned}\angle MPR &= \angle LPM - \angle LPS \\ &= \angle LPM - \angle KLM \\ &= \angle LPM - \angle KPM \\ &= \angle LPK = \angle LMK = \angle SMR.\end{aligned}$$

Therefore triangles RSM and RMP are similar; then

$$RM/RP = RS/RM \text{ and } RM^2 = RS \cdot RP.$$

We already have $RN^2 = RS \cdot RP$ since RN is tangent to C, so we conclude that R is the midpoint of MN and the circumcenter of MNQ. (Also, since $\angle MPR = \angle SMR$, the circumcircle of SMP is tangent to NM at M. So MN is tangent to circumcircles NPS and MPS. Then line PS is the radical axis of the two circumcircles. Since R is on line PS, R has equal power to the two circles and $RN^2 = RM^2$.)

Now we have $\angle MPR = \angle RMS = \angle MQR$, so M, R, P, Q are concyclic and

$$\angle MPQ = \angle MPS + \angle SPQ = \angle KML + \angle KML = 2\angle KML.$$

Second Solution. Invert about L. We will find that the diagram inverts such that LK' is parallel to $P'N'$ and $P'N'$ is tangent to the circumcircle of L, K', and N' at N'. Furthermore, M' is the intersection of $K'P'$ and the circumcircle of L, K', and N'. Also, Q' lies on LM' such that $\angle LN'Q'$ is a right angle. Now let us construct point P_1, the reflection of P' across N'. Let P_2 be the intersection of LP_1 and $Q'N'$ and O be the center of the circumcircle. Let $\phi = \angle K'LP_1$ and $\theta = \angle P_2N'P_1$. Since $LK' \parallel P'P_1$, $\angle LP_1P' = \phi$. By symmetry, $\angle K'P'P_1 = \phi$ as well.

Now $\angle LP_2N' = \phi + \theta$. Since $P'N'$ is tangent to the circumcircle at N', $ON' \perp P'N'$. So, $\angle ON'L = \theta$ and by symmetry, $\angle ON'K' = \theta$. Thus, $\angle LM'K' = 2\theta$. Let I be the intersection of LP_1, $K'P'$, and ON'. Now $\angle LIP' = 2\phi$. Since the sum of the angles in a triangle is $180°$, $\angle ILM' = 180° - 2\theta - 2\phi$. Now we can find the angle measure of $\angle LQ'P_2$ to be $\theta+\phi$, and triangle LP_2Q' is isosceles. Since $LN' \perp P_2Q'$, $Q'N' = N'P_2$. By vertical angles, $\angle P_1N'P_2 = \angle P'N'Q'$. Finally $N'P' = N'P_1$, so $\triangle P_1P_2N' \cong \triangle P'Q'N'$. So, $\angle N'P'Q' = \phi$, and $\angle M'P'Q' = 2\phi = 2\angle K'LP_1 = 2\angle M'K'L$. From inversive angles, we discover that $\angle MPQ = 2\angle KML$.

Problem 2

Suppose an $n \times n$ table is filled with the numbers 0, 1, -1 in such a way that every row and column contains exactly one 1 and one -1. Prove that the rows and columns can be reordered so that in the resulting table each number has been replaced by its negative.

Solution. Construct a directed graph whose vertices are the rows of the table, with one edge e_i from the row in which column i has $+1$ to the row in which it has -1. This graph has in-degree and out-degree 1 at every vertex, so it is the disjoint union of cycles. A suitable permutation of rows reverses all of the cycles; by permuting the columns to get nonzero entries in the original nonzero positions, we get the desired arrangement.

Problem 3

Let x_1, x_2, x_3, x_4 be positive real numbers such that

$$x_1x_2x_3x_4 = 1.$$

Prove that

$$\sum_{i=1}^{4} x_i^3 \geq \max\left\{\sum_{i=1}^{4} x_i, \sum_{i=1}^{4} \frac{1}{x_i}\right\}.$$

Solution. Let $A = \sum x_i^3$ and $A_i = A - x_i^3$, so that $A = \frac{1}{3}\sum A_i$. We claim that $A \geq \sum \frac{1}{x_i}$ and $A \geq \sum x_i$. From AM-GM,

$$\frac{1}{3}A_1 \geq \sqrt[3]{x_2^3x_3^3x_4^3} = \frac{1}{x_1}.$$

Combining the analogous inequalities gives $A \geq \sum \frac{1}{x_i}$, as claimed.

Also, by the power mean inequality,

$$\frac{1}{4}A \geq \left(\frac{\sum x_i}{4}\right)^3 \geq \left(\frac{\sum x_i}{4}\right)\left(\frac{\sum x_i}{4}\right)^2 \geq \frac{\sum x_i}{4},$$

since $\sum x_i \geq 4$ by AM-GM. So $A \geq \sum x_i$, as claimed.

Problem 4

Let ABC be an acute triangle and let D be the foot of the altitude from A. Let the angle bisectors of B and C meet AD at E and F, respectively. If $BE = CF$, prove that ABC is an isosceles triangle.

First Solution. We claim that $AB = AC$. Suppose on the contrary that $AB < AC$. Then $BD < CD$. Let B' be the reflection of B across D, which lies between C and D.

If E were between D and F, we would have $BE = B'E < CE \leq CF$, contradiction. Thus E is between A and F. In particular, the segments EB' and FC meet at a point X. Since X lies on the angle bisector of $\angle BCA$, X is equidistant to AC and BC. Also, X lies on the angle bisector of $\angle AB'D$, so it is equidistant to AB' and DC. But this is impossible, since the perpendicular dropped from X to AC meets AB' before it meets AC. Contradiction. So our assumption is wrong and $AB = AC$.

Second Solution. We claim that $AB = AC$. Suppose on the contrary that $AB < AC$. Then $\angle ACB < \angle ABC$. Since ABC is acute, $\cos\angle ACB > \cos\angle ABC$. From angle bisector theorem,

$$\frac{DF}{FA} = \frac{DC}{CA} = \cos\angle ABC > \cos\angle ABC = \frac{DB}{BA} = \frac{DE}{EA}.$$

Thus E is on FD. In right triangle BDF, $BE < BF$. Since $AB < AC$, line AD is closer to B than to C and $BF < CF$. Thus $BE < BF < CF$, a contradiction. So our assumption is wrong and $AB = AC$.

Third Solution. Let $\theta = \angle FCD$ and $\phi = \angle EBD$. Let $CF = 1 = BE$. By trigonometry, $CD = \cos\theta$ and thus $AD = \tan 2\theta \cos\theta$. Similarly, $AD = \tan 2\phi \cos\phi$. For $0 < x < \pi/2$, $f(x) = \tan 2x \cos x$ is a monotonic increasing function as

$$\begin{aligned} f'(x) &= 2\sec^2 2x \cos x - \tan 2x \sin x \\ &= \frac{2\cos x(1 - \sin^2 x \cos 2x)}{\cos^2 2x} > 0. \end{aligned}$$

Since $BE = CF$, $\theta = \phi$ and $AB = AC$.

Problem 5

Suppose a, b are natural numbers such that

$$p = \frac{b}{4}\sqrt{\frac{2a-b}{2a+b}}$$

is a prime number. What is the maximum possible value of p?

Solution. The largest p is 5. Note that b is even, so we may write $b = 2c$. Now

$$p = \frac{c}{2}\sqrt{\frac{a-c}{a+c}} \text{ or } \frac{4p^2}{c^2} = \frac{a-c}{a+c}.$$

Write $2p/c = m/n$ in lowest terms. If $k = \gcd(a-c, a+c)$, then $a - c = km^2, a + c = kn^2$, so $2c = k(n^2 - m^2)$ and

$$4pn = km(n^2 - m^2).$$

In case m, n are both odd, then $8|m^2 - n^2$ and so p is even, that is, $p = 2$. On the other hand, if m and n are not both odd, $n^2 - m^2$ is odd, so k must be even. Write $k = 2r$, so that we have $2pn = rm(n^2 - m^2)$. Now n is coprime to m and to $n^2 - m^2$, so n divides r. Writing $r = ns$, we have $2p = s(n-m)(n+m)m$.

Suppose $p > 2$. Since $m + n$ and $n - m$ are odd, $m + n = p$ and $n - m = 1$. Thus s, m are each at most 2. This leaves only the possibilities $(m, n) = (1, 2)$ or $(2, 3)$. In either case, $p \leq 5$, and fortunately this can be achieved with $m = 2$, $n = 3$, $s = 1$, $r = 3$, $k = 6$, $c = 15$, $b = 30$, and $a = 39$.

Problem 6

Let x, a, b be positive integers such that $x^{a+b} = a^b b$. Prove that $a = x$ and $b = x^x$.

Solution. If $x = 1$, then $a = b = 1$ and we are done. So we may assume $x > 1$. Write $x = \prod_{i=1}^{n} p_i^{\gamma_i}$, where the p_i are the distinct prime factors of x. Since a and b divide x^{a+b}, we have $a = \prod p_i^{\alpha_i}$ and $b = \prod p_i^{\beta_i}$ for some nonnegative integers α_i, β_i.

First suppose that some β_i is zero, that is, p_i does not divide b. Then the given equation implies that $\gamma_i(a+b) = \alpha_i b$, so that $(\alpha_i - \gamma_i)b = a\gamma_i$. Now $p_i^{\alpha_i}$ divides a but is coprime to b, so $p_i^{\alpha_i}$ divides $\alpha_i - \gamma_i$ also. But $p_i^{\alpha_i} > \alpha_i$ for $\alpha_i > 0$, contradiction. We conclude that $\beta_i > 0$.

Now from the fact that

$$\gamma_i(a+b) = \beta_i + b\alpha_i$$

and the fact that p^{β_i} does not divide β_i (again for size reasons), we deduce that p^{β_i} also does not divide a, that is, $\alpha_i < \beta_i$ for all i and so a divides b. Moreover, the equation above implies that a divides β_i, so we may write $b = c^a$ with $c \geq 2$ a positive integer.

Write $x/a = p/q$ in lowest terms (so $\gcd(p,q) = 1$). Then the original equation becomes $x^a p^b = bq^b$. Now p^b must divide b, which can only occur if $p = 1$. That is, x divides a.

If $x \neq a$, then there exists i with $\alpha_i \geq \gamma_i + 1$, so

$$\gamma_i(a+b) = \beta_i + \alpha_i b \geq (\gamma_i + 1)b$$

and so $\gamma_i a > b$. On the other hand, a is divisible by $p_i^{\gamma_i}$, so in particular $a \geq \gamma_i$. Thus $a^2 > b = c^a$, or $\sqrt{c} < a^{1/a}$; however, $a^{1/a} < \sqrt{2}$ for $a \geq 5$, so this can only hold for $c = 2$ and $a = 3$, in which case $b = 8$ is not divisible by a, contrary to our earlier observation.

Thus $x = a$, and from the original equation we get $b = x^x$, as desired.

Problem 7

Let ABC be an acute triangle and D, E, F the feet of its altitudes from A, B, C, respectively. The line through D parallel to EF meets line AC and line AB at Q and R, respectively. Let P be the intersection of line BC and line EF. Prove that the circumcircle of PQR passes through the midpoint of BC.

First Solution. Let M be the midpoint of BC. From the assumption $EF \parallel RQ$, we conclude that

$$\angle BRQ = \angle AFE = \angle ACB = \angle QCB$$

and so quarilateral $BRQC$ is cyclic. Therefore $BD \cdot DC = RD \cdot DQ$. On the other hand, AD, BE, CF are concurrent cevians, so that D and P are harmonic conjugates with respect to B and C. (In fact, from Ceva's theorem, we have $CE \cdot AF \cdot BD = EA \cdot FB \cdot DC$; from Menelaus's theorem, $AF \cdot BP \cdot CE = FB \cdot PC \cdot EA$. Dividing them leads to $DB/BP = DC/CP$). Therefore

$$MB^2 = MP \cdot MD = MD(MD + PD) = MD^2 + MD \cdot PD$$

and so $MD \cdot PD = MB^2 - MD^2$, i.e.,

$$MD \cdot PD = (MB - MD)(MB + MD) = BD \cdot CD = RD \cdot QD.$$

Thus P, Q, R, M are concyclic, as desired.

Second Solution. We use directed lengths. We know that $EFMD$ is cyclic (9 point circle). Since $\angle CEB = 90° = \angle CFB$, $EFBC$ is cyclic. So line EF is the radical axis of the circumcircles $EFMD$ and $EFBC$. Since P is on EF, P has equal power to the two circles and

$$\begin{aligned}
PC \cdot PB &= PD \cdot PM,\\
PC(PM + MB) &= (PC + CD)PM,\\
PC \cdot MB &= CD \cdot PM,\\
PC(CD + DM) &= CD(PC + CM),\\
CD \cdot CM &= PC \cdot DM,\\
CD \cdot CM + CD \cdot DM &= PC \cdot DM + CD \cdot DM,\\
CD \cdot MB + CD \cdot DM &= PD \cdot DM,\\
CD \cdot DB &= PD \cdot DM.
\end{aligned}$$

As in the first solution, we know that $BRCQ$ is cyclic and $DB \cdot DC = DR \cdot DQ$. Thus $PD \cdot DM = RD \cdot DQ$ and P, Q, R, M are concyclic, as desired.

Third Solution. Let a, b, c and A, B, C denote the sides and angle measures of triangle ABC, respectively. WLOG, assume that $\angle C \geq \angle B$. Observe that

$$\frac{c}{b}\cos B - \cos C = \frac{\sin C \cos B}{\sin B} - \frac{\cos C \sin B}{\sin B} = \frac{\sin(C - B)}{\sin B}.$$

And $\angle BDR = \angle P = \angle AFE - \angle B = C - B$. So,

$$\frac{\sin(C - B)}{\sin B} = \frac{\sin \angle BDR}{\sin B} = \frac{BR}{RD}.$$

But

$$\begin{aligned}
DM = CM - CD &= \frac{1}{2}CB - CA\cos C\\
&= \frac{1}{2}(b\cos C + c\cos B) - b\cos C\\
&= \frac{1}{2}(c\cos B - b\cos C).
\end{aligned}$$

And

$$2\frac{DM}{CA} = \frac{c}{b}\cos B - b\cos C = \frac{BR}{RD} = \frac{CE}{PC}$$

$$\Longrightarrow PC \cdot DM = \frac{CE \cdot CA}{2} = \frac{CD \cdot CB}{2} = CD \cdot MB$$

$$\Longrightarrow CD \cdot DM + PC \cdot DM = CD \cdot MB + CD \cdot DM$$

$$\Longrightarrow PD \cdot DM = CD \cdot DB.$$

As in the first solution, we know that $BRCQ$ is cyclic and $DB \cdot DC = DR \cdot DQ$. Thus $PD \cdot DM = RD \cdot DQ$ and P, Q, R, M are concyclic, as desired.

Problem 8

Let $S = \{x_0, x_1, \ldots, x_n\} \subset [0, 1]$ be a finite set of real numbers with $x_0 = 0$, $x_1 = 1$, such that every distance between pairs of elements occurs at least twice, except for the distance 1. Prove that all of the x_i are rational.

Solution. The set S spans some finite-dimensional vector space over $\mathbb{Q}$; let $\beta_1, \ldots, \beta_m$ be a basis of this vector space with $\beta_m = 1$. For each i, we write

$$x_i = q_{i1}\beta_1 + \cdots + q_{im}\beta_m$$

and define the vector $v_i = (q_{i1}, \ldots, q_{im})$.

Let us compare the v_i in lexicographic order: $v_i > v_j$ if in the first position where they differ, v_i has the larger component. Let v_s and v_t be the largest and smallest of the v_i, respectively, and suppose a and b are such that $x_s - x_t = x_a - x_b$; since the β_k form a basis, this also implies $v_s - v_t = v_a - v_b$. By our choice of v_s and v_t, this can only happen if $a = s$ and $b = t$. Thus for $0 < i < n$,

$$v_s = (0, 0, \ldots, 1) > v_i > v_t = (0, 0, \ldots, 0).$$

This is only possible if the v_i only have nonzero entries in the last component, which is to say the x_i are all rational.

Problem 9

Let $x, y, z > 1$ and $\frac{1}{x} + \frac{1}{y} + \frac{1}{z} = 2$. Prove that

$$\sqrt{x+y+z} \geq \sqrt{x-1} + \sqrt{y-1} + \sqrt{z-1}.$$

Solution. By Cauchy-Schwarz Inequality,

$$\sqrt{x+y+z}\sqrt{\frac{x-1}{x}+\frac{y-1}{y}+\frac{z-1}{z}} \geq \sqrt{x-1}+\sqrt{y-1}+\sqrt{z-1}.$$

On the other hand, by hypothesis,

$$\frac{x-1}{x}+\frac{y-1}{y}+\frac{z-1}{z}=1,$$

so the desired result follows.

Problem 10

Let P be the set of all points in $\mathbb{R}^n$ with rational coordinates. For $A, B \in P$, one can move from A to B if the distance AB is 1. Prove that every point in P can be reached from every other point in P by a finite sequence of moves if and only if $n \geq 5$.

Solution. Note that if any two points in P can be connected for a given n, then the same is true for any larger n. Thus it suffices to show that this is not the case for $n = 4$ and that it is the case for $n = 5$.

First suppose $n = 4$. We say that a rational number is a 2-integer if it can be written as p/q with p, q integers and q odd. We will show that for any point (x_1, x_2, x_3, x_4) in P reachable from the origin, $x_1^2+x_2^2+x_3^2+x_4^2$ is a 2-integer. Thus, for example, $(1/2, 0, 0, 0)$ will not be reachable, and the assertion of the problem for $n = 4$ will follow.

We proceed by induction on the minimum number of steps required to reach the point. Obviously the claim holds for the origin. Now suppose the claim holds for (x_1, x_2, x_3, x_4) and (y_1, y_2, y_3, y_4) is a unit vector with rational components. Write $x_i = p_i/q_i$ and $\sum x_i^2 = p/q$ in lowest terms; then p/q is a 2-integer and q is odd. We also notice that the q_i must all be odd or must all be twice an odd number. If the least common multiple d of the q_i is not odd, then $\sum(p_i d/q_i)^2 = d^2p/q$ is an integer. Since d is even and q is odd, the right side is 0 mod 4, while the left side is the sum of terms which are 0 or 1 mod 4, and not all of the former. Hence, they are all of the latter form, so the q_i are all divisible by exactly the same power of 2. Moreover, the left side is now 4 mod 8, so d is not divisible by 4.

Likewise, if we write $y_i = v_i/w_i$ in lowest terms ($\sum y_i^2 = 1$), the w_i are either all odd or all twice an odd number. Let e be the least common multiple of the w_i. Then

$$\sum(x_i+y_i)^2=\sum x_i^2+\sum y_i^2+\frac{2}{de}\sum(dx_i)(ey_i).$$

The first two sums on the right side are 2-integers by the induction hypothesis. Moreover, the third sum is the sum of 4 odd integers, multiplied by 2 and divided by an integer not divisible by 8. Therefore it is also a 2-integer. Since the sum of 2-integers is a 2-integer, we have that $\sum(x_i+y_i)^2$ is a 2-integer, and the induction is complete.

Now suppose $n \geq 5$; in fact, it suffices to consider $n=5$. The key fact is that any positive integer can be written as the sum of four perfect squares (Legendre's theorem). From this it follows that any rational number is the sum of four rational perfect squares.

It suffices to show that one can get from the origin to the point $(a,0,0,0,0)$. To do so, find a solution to

$$(a/2)^2+x_1^2+x_2^2+x_3^2+x_4^2=1$$

in rational numbers x_1, x_2, x_3, x_4, and then step from the origin to $(a/2, x_1, x_2, x_3, x_4)$ to $(a,0,0,0,0)$.

Problem 11

Let $f_1, f_2, f_3 : \mathbb{R} \to \mathbb{R}$ be functions such that

$$a_1f_1+a_2f_2+a_3f_3$$

is monotonic for all $a_1, a_2, a_3 \in \mathbb{R}$. Prove that there exist $c_1, c_2, c_3 \in \mathbb{R}$, not all zero, such that

$$c_1f_1(x)+c_2f_2(x)+c_3f_3(x)=0$$

for all $x \in \mathbb{R}$.

First Solution. We establish the following lemma.

Lemma 11. *Let $f, g : \mathbb{R} \to \mathbb{R}$ be functions such that f is nonconstant and $af+bg$ is monotonic for all $a, b \in \mathbb{R}$. Then there exists $c \in \mathbb{R}$ such that $g-cf$ is a constant function.*

Proof. Let s, t be two real numbers such that $f(s) \neq f(t)$. Let $u=(g(s)-g(t))/(f(s)-f(t))$. Let $h_1=g-a_1f$ for some $a_1 \in R$. Then h_1 is monotonic. But

$$h_1(s)-h_1(t)=g(s)-g(t)-a_1(f(s)-f(t))=(f(s)-f(t))(u-a_1).$$

Since $f(s) - f(t) \neq 0$ is fixed, the monotonicity of h_1 depends only on the sign of $u - a_1$.

Since f is nonconstant, there exist $x_1, x_2 \in \mathbb{R}$ such that $f(x_1) \neq f(x_2)$. Let

$$c = \frac{g(x_1) - g(x_2)}{f(x_1) - f(x_2)}$$

and $h = g - cf$. Then $r = h(x_1) = h(x_2)$ and the monotonicity of $h_1 = g - a_1 f$, for each a_1, depends only on the sign of $c - a_1$. We claim that $h = g - cf$ is a constant function.

We prove our claim by contradiction. Suppose, on the contrary, that there exists $x_3 \in \mathbb{R}$ such that $h(x_3) \neq r$. Since $f(x_1) \neq f(x_2)$, at least one of $f(x_1) \neq f(x_3)$ and $f(x_2) \neq f(x_3)$ is true. WLOG, suppose that $f(x_1) \neq f(x_3)$. Let

$$c' = \frac{g(x_1) - g(x_3)}{f(x_1) - f(x_3)}.$$

Then the monotonicity of h_1 also depends only on the sign of $c' - a_1$.

Since $h(x_3) \neq r = h(x_1)$,

$$c \neq \frac{g(x_1) - g(x_3)}{f(x_1) - f(x_3)} = c'$$

and hence $c - a_1 \neq c' - a_1$. So there exists some a_1 such that h_1 is both strictly increasing and decreasing, which is impossible. Therefore our assumption is false and h is a constant function. $\square$

Now we prove our main result. If f_1, f_2, f_3 are all constant functions, the result is trivial. WLOG, suppose that f_1 is nonconstant. For $a_3 = 0$, we apply the lemma to f_1 and f_2, so $f_2 = cf_1 + d$; for $a_2 = 0$, we apply the lemma to f_1 and f_3, so $f_3 = c' f_1 + d'$. Here c, c', d, d' are constant. We have

$$(c'd - cd')f_1 + d' f_2 - df_3 = (c'd - cd')f_1 + d'(cf_1 + d) - d(c' f_1 + d') = 0.$$

If $(c'd - cd', d', -d) \neq (0, 0, 0)$, then let

$$(c_1, c_2, c_3) = (c'd - cd', d', -d)$$

and we are done. Otherwise, $d = d' = 0$ and f_2, f_3 are constant multiples of f_1. Then the problem is again trivial.

Second Solution. Define the vector

$$v(x) = (f_1(x), f_2(x), f_3(x))$$

for $x \in \mathbb{R}$. If the $v(x)$ span a proper subspace of $\mathbb{R}^3$, we can find a vector (c_1, c_2, c_3) orthogonal to that subspace, and then $c_1 f_1(x) + c_2 f_2(x) + c_3 f_3(x) = 0$ for all $x \in \mathbb{R}$.

So suppose the $v(x)$ span all of $\mathbb{R}^3$. Then there exist $x_1 < x_2 < x_3 \in \mathbb{R}$ such that $v_1 = v(x_1)$, $v_2 = v(x_2)$, $v_3 = v(x_3)$ are linearly independent, and so the 3×3 matrix A with $A_{ij} = f_j(x_i)$ has linearly independent rows. But then A is invertible, and its columns also span $\mathbb{R}^3$. This means we can find c_1, c_2, c_3 with $c_1 v_1 + c_2 v_2 + c_3 v_3 = (0, 1, 0)$, and the function $c_1 f_1 + c_2 f_2 + c_3 f_3$ is then not monotonic, a contradiction.

Problem 12

Let X be a finite set with $|X| = n$ and let $A_1, A_2, \cdots, A_m$ be three-element subsets of X such that $|A_i \cap A_j| \leq 1$ for all $i \neq j$. Show that there exists a subset A of X with at least $\sqrt{2n}$ elements containing none of the A_i.

Solution. Let A be a subset of X containing no A_i, and having the maximum number of elements subject to this condition. Let k be the size of A. By assumption, for each $x \in X - A$, there exists $i(x) \in \{1, \ldots, m\}$ such that $A_{i(x)} \subseteq A \cup \{x\}$. Let $L_x = A \cap A_{i(x)}$, which by the previous observation must have 2 elements. Since $|A_i \cap A_j| \leq 1$ for $i \neq j$, the L_x must all be distinct. From this we conclude $n - k \leq \binom{k}{2}$, and so $k^2 + k \geq 2n$, and so $k \geq \lfloor \sqrt{2n} \rfloor$.

Problem 13

The edges of a regular 2^n-gon are colored red and blue in some fashion. A step consists of recoloring each edge whose neighbors are both the same color in red, and recoloring each edge whose neighbors are of opposite colors in blue. Prove that after 2^{n-1} steps all of the edges will be red, and show that this need not hold after fewer steps.

First Solution. Number the edges $1, \ldots, 2^n$ around the polygon, and let $a_i^{(j)}$ be 1 if edge i is red after j steps and -1 if edge i is blue after j steps. Then $a_i^{(j)} = a_{i-1}^{(j-1)} a_{i+1}^{(j-1)}$, where $a_{2^n+1}^{(j)} = a_1^{(j)}$ and $a_0^{(j)} = a_{2^n}^{(j)}$ for all j. We need to prove that all $a_i^{(2^{n-1})} = 1$ for all i. We prove our result inductively. It is easy to check that the statement is true for $n = 2$.

Now, suppose $n \geq 2$ and the statement holds for all smaller values of n. We observe the first two steps:

$$(a_1^{(1)}, a_2^{(1)}, \ldots, a_{2^n}^{(1)}) = (a_1, a_2, a_3, a_4, a_5, \ldots, a_{2^n})$$

$$
\begin{aligned}
&\longrightarrow \quad (a_1^{(2)}, a_2^{(2)}, \ldots, a_{2^n}^{(2)}) \\
&\quad = (a_2 a_{2^n}, a_1 a_3, a_2 a_4, a_3 a_5, \ldots, a_{2^n-1} a_1) \\
&\longrightarrow \quad (a_1^{(3)}, a_2^{(3)}, \ldots, a_{2^n}^{(3)}) \\
&\quad = (a_3 a_{2^n-1}, a_4 a_{2^n}, a_1 a_5, \ldots, a_2 a_{2^n-2})
\end{aligned}
$$

since $a_i^2 = 1$. If we only consider all the odd numbered edges in the above, we have

$$(a_1, a_3, \ldots, a_{2^n-1}) \longrightarrow (a_3 a_{2^n-1}, a_1 a_5, \ldots, a_{2^n-3} a_1)$$

in two steps, which is equivalent to a one step move for a 2^{n-1}-gon with starting value $(a_1, a_3, a_5, \ldots, a_{2^n-1})$. We have a similar result for all even edges of the 2^n-gon. From the above we see that a two-step move in a 2^n-gon is equivalent to a one-step move in each of two (odd and even) sub 2^{n-1}-gon's. Our result follows from the inductive hypothesis.

Second Solution. Number the edges $1, \ldots, 2^n$ around the polygon, and let $x_i^{(j)}$ be 0 if edge i is red after j steps and 1 if edge i is blue after j steps. Then there is a matrix A such that

$$\begin{bmatrix} x_1^{(j+1)} \\ \vdots \\ x_{2^k}^{(j+1)} \end{bmatrix} \equiv A \begin{bmatrix} x_1^{(j)} \\ \vdots \\ x_{2^k}^{(j)} \end{bmatrix} \pmod 2.$$

To be precise, $A = P + P^{-1}$, where P is the matrix with $P_{ij} = 1$ if $j - i \equiv 1 \pmod{2^n}$ and 0 otherwise. Note that P^r is the matrix with $(P^r)_{ij} = 1$ if $j - i \equiv r \pmod{2^n}$ and 0 otherwise.

Now recall that for any matrices A and B, $(A+B)^2 \equiv A^2 + B^2 \pmod 2$, and so by induction, $(A+B)^{2^m} \equiv A^{2^m} + B^{2^m} \pmod 2$. In particular,

$$A^{2^{n-1}} \equiv P^{2^{n-1}} + P^{-2^{n-1}} = 2P^{2^{n-1}} \equiv 0 \pmod 2.$$

Thus after 2^{n-1} steps, all edges are red.

If we start with exactly one blue edge, it is easily seen that after m steps ($m < 2^{n-1}$), the edge m away from the initially blue edge is now blue, and so no fewer than 2^{n-1} steps will make all of the edges red.

Problem 14

Let $n_1 < n_2 < \cdots$ be a sequence of natural numbers such that for $i < j$, the decimal representation of n_i does not occur as the leftmost digits of the decimal representation of n_j. (For example, 137 and 13729 cannot both occur in the sequence.) Prove that

$$\sum_{i=1}^{\infty} \frac{1}{n_i} \leq 1 + \frac{1}{2} + \cdots + \frac{1}{9}.$$

Solution. Clearly it suffices to prove the claim for each finite sequence. Suppose a finite sequence is given, and let $M = 10N + d$ be the largest element of the sequence, with $0 \leq d \leq 9$. Then N does not belong to the sequence. Moreover, removing $10N, 10N+1, \ldots, 10N+9$ from the sequence if they occur and adding N gives another sequence whose sum of reciprocals is

$$\sum_i \frac{1}{n_i} + \frac{1}{N} - \sum_{i=0}^{9} \frac{1}{10N+i} \geq \sum_i \frac{1}{n_i}.$$

Thus we can repeatedly make such substitutions and never decrease the sum of reciprocals. This process must terminate(since the sequence is finite), and only does so when the sequence is $\{1, \ldots, 9\}$, so this sequence has the largest sum of reciprocals.

Problem 15

Let K be a convex polygon in the plane. Prove that for any triangle containing K of minimum area, the midpoints of its sides lie on K.

Solution. Suppose PQR is a triangle containing K but the midpoint G of QR does not lie on K. We will show that PQR is not a triangle of minimal area containing K. Let ℓ be a line separating K from G (which exists because K is convex). Let O be the intersection of ℓ with QR. Line ℓ intersects either PQ or PR. WLOG, let ℓ intersect PQ at N. Reflect triangle ONQ about O to obtain a new triangle $ON'Q'$, and extend PR to meet one of ON' or $Q'N'$.

If PR meets side ON' first at M', then

$$[OM'R] < [ON'Q'] = [ONQ]$$

and so $PM'N$ is a triangle containing K with smaller area than PQR. Otherwise, let PR meet side $Q'N'$ at M', and let M be the reflection of

M' across O; then

$$[OM'R] < [OM'Q'] = [OMQ]$$

and triangle PMM' contains K and has smaller area than PQR.

Problem 16

Let ABC be a triangle. Extend the side BC past C, and let D be the point on the extension such that $CD = AC$. Let P be the second intersection of the circumcircle of ACD with the circle with diameter BC. Let BP and AC meet at E, and let CP and AB meet at F. Prove that D, E, F are collinear.

Solution. Let AP meet BC at D'. By Ceva's theorem,

$$\frac{AF}{FB}\frac{BD'}{D'C}\frac{CE}{EA} = 1$$

and so by Menelaus's theorem, it suffices to show that

$$\frac{DB}{DC} = \frac{BD'}{D'C}$$

(as signed ratios) to conclude that D, E, F are collinear. (In other words, we must show that D and D' are harmonic conjugates with respect to B and C.)

Since $AC = AD$, we have $\angle DPA = 180° - 2\angle DPC$, and so $\angle DPC = \angle CPD'$, which is to say PC is the internal angle bisector of $\angle DPD'$. Since $PB \perp PC$, PB is the external angle bisector of $\angle DPD'$. From the (internal and external) angle bisector theorem,

$$\frac{DB}{D'B} = \frac{DP}{D'P} = \frac{DC}{D'C}$$

as desired. (In fact, the circle with diameter BC is the circle of Apollonius of $D'D$.)

1.8 Ireland

Problem 1

Show that if x is a nonzero real number, then

$$x^8 - x^5 - \frac{1}{x} + \frac{1}{x^4} \geq 0.$$

Solution. We have

$$x^8 - x^5 - \frac{1}{x} + \frac{1}{x^4} = x^5(x^3-1) - \frac{x^3-1}{x^4} = \frac{(x^9-1)(x^3-1)}{x^4}.$$

Since $x^4 \geq 0$, and $x^9 - 1$ and $x^3 - 1$ have the same sign, the expression is always positive or zero.

Problem 2

The point P lies inside an equilateral triangle and its distances to the three vertices are 3, 4, 5. Find the area of the triangle.

Solution. Label the vertices of the equilateral triangle A, B, C such that $PA = 3$, $PB = 4$, $PC = 5$. Rotate triangle BPA 60 degrees around A so that the edges AB and AC coincide. Let X be the image of point P. Observe that $PX = 3$, $CX = 4$, and $CP = 5$; thus, $PX \perp CX$. Also, APX becomes an equilateral triangle of side 3. Now, applying the law of cosines to triangle CXA, we find

$$AC = \sqrt{3^2 + 4^2 - 2 \cdot 3 \cdot 4 \cos 150^\circ} = \sqrt{25 + 12\sqrt{3}},$$

and the area of the triangle is $25\sqrt{3}/4 + 9$.

Problem 3

Show that no integer of the form $xyxy$ in base 10 can be the cube of an integer. Also find the smallest base $b > 1$ in which there is a perfect cube of the form $xyxy$.

Solution. If the 4-digit number $xyxy = 101 \times xy$ is a cube, then $101 | xy$, which is a contradiction.

Convert $xyxy = 101 \times xy$ from base b to base 10. We find $xyxy = (b^2+1) \times (bx + y)$ with $x, y < b$ and $b^2 + 1 > bx + y$. Thus for $xyxy$ to be a cube, $b^2 + 1$ must be divisible by a perfect square. We can check easily that $b = 7$ is the smallest such number, with $b^2 + 1 = 50$. The smallest cube divisible by 50 is 1000 which is 2626 in base 7.

Problem 4

Show that a disc of radius 2 can be covered by seven (possibly overlapping) discs of radius 1.

Solution. We pack 7 congruent regular hexagons, with side length 1, in a honeycomb shape—1 hexagon in the middle with each of its sides attached to another hexagon. It is easy to see that the resulting shape can cover a circle of radius 2. Since we can cover each one of the hexagons with a unit circle, we are done.

Problem 5

If x is a real number such that $x^2 - x$ is an integer, and for some $n \geq 3$, $x^n - x$ is also an integer, prove that x is an integer.

First Solution. Let integer $z = x^2 - x$; solve the quadratic equation $x^2 - x - z = 0$ for x. We have

$$x = \frac{1 \pm \sqrt{a}}{2} \quad \text{with } a = 1 + 4z$$

Since $x^n - x = x(x-1)(x^{n-2} + x^{n-3} + x^{n-4} + \cdots + 1)$, we know that

$$x^{n-2} + x^{n-3} + x^{n-4} + \cdots + 1 = \sum_{i=0}^{n-2} x^i = \frac{x^n - x}{x^2 - x}$$

is rational. We claim that x is rational. We prove our claim by contradiction. Suppose that x is irrational. We consider the following two cases:

(a) $x = (1 + \sqrt{a})/2$. Then all powers of x will be of the form $\alpha + \beta\sqrt{a}$, where α, β are positive rationals. So, $\sum_{i=0}^{n-2} x^i = \alpha' + \beta'\sqrt{a}$, where α', β' are also positive rationals and the sum is irrational.

(b) $x = (1 - \sqrt{a})/2$. Then all powers of x will be of the form $\alpha + \beta\sqrt{a}$ where α is a positive rational and β is a negative rational. This is because the square root is preserved if and only if $-\sqrt{a}$ is raised to an odd power—this preserves the negative sign as well. Thus, the sum will be $\alpha' - \beta'\sqrt{a}$, where β' is a positive rational and α' is a negative rational. And again, the sum is irrational.

Both cases leads to a contradiction. Therefore, x must be rational; let $x = p/q$ with $\gcd(p, q) = 1$. Thus

$$z = x^2 - x = \frac{p(p-q)}{q^2}$$

However, $\gcd(p,q) = 1$ implies that

$$\gcd(p(p-q),q) = \gcd(p-q,q) = \gcd(p,q) = 1.$$

Therefore, z is an integer only if $q = 1$, which implies that $x = p$ is an integer.

Second Solution. Under the same notation, we claim that $x^n - x = kx + mz$ for some integers k and m, which imples that $x = (x^n - x - ma)/k$ is rational and the rest is the same. Let l be a positive integer. We notice that

(a) $x^{2l} = (x^2)^l = (x+z)^l$;
(b) $x^{2l+1} = x(x^2)^l = x(x+z)^l$.

It easy to see that the above operations reduce the power of x and operate with integers only. Repeated application of the above operations easily leads to our claim.

Problem 6

Find all positive integers n that have exactly 16 positive integral divisors $d_1, d_2, \ldots, d_{16}$ such that

$$1 = d_1 < d_2 < \cdots < d_{16} = n,$$

$d_6 = 18$ and $d_9 - d_8 = 17$.

Solution. Let integer $n = p_1^{a_1} p_2^{a_2} \cdots p_m^{a_m}$ with $p_1, \ldots, p_m$ distinct primes. Then n has $(a_1+1)(a_2+1)\cdots(a_n+1)$ divisors. Since $18 = 2 \cdot 3^2$, it has 6 factors: 1, 2, 3, 6, 9, 18. Since d has 16 divisors, we know that $d = 2 \cdot 3^3 \cdot p$ or $d = 2 \cdot 3^7$. If $b = 2 \cdot 3^7$, $d_8 = 54$, $d_9 = 81$ and $d_9 - d_8 \neq 17$. Thus $d = 2 \cdot 3^3 \cdot p$ for some prime $p > 18$. If $p < 27$, then $d_7 = p, d_8 = 27, d_9 = 2p = 27 + 17 = 44 \Longrightarrow p = 22$, a contradiction. Thus $p > 27$. If $p < 54$, $d_7 = 27, d_8 = p, d_9 = 54 = d_8 + 17 \Longrightarrow p = 37$. If $p > 54$, then $d_7 = 27, d_8 = 54, d_9 = d_8 + 17 = 71$. We obtain two solutions for the problem: $2 \cdot 3^3 \cdot 37 = 1998$ and $2 \cdot 3^3 \cdot 71 = 3834$.

Problem 7

Prove that if a, b, c are positive real numbers, then

$$\frac{9}{a+b+c} \leq 2\left(\frac{1}{a+b} + \frac{1}{b+c} + \frac{1}{c+a}\right)$$

and
$$\frac{1}{a+b}+\frac{1}{b+c}+\frac{1}{c+a}\le\frac{1}{2}\left(\frac{1}{a}+\frac{1}{b}+\frac{1}{c}\right).$$

Solution. The first inequality can be directly concluded from either AM-HM or Cauchy-Schwarz Inequality, as it can be rewritten as
$$\frac{9}{(a+b)+(b+c)+(c+a)}\le\frac{1}{a+b}+\frac{1}{b+c}+\frac{1}{c+a}.$$

We can also use Jensen's inequality to prove both inequalities. We know that $f(x)=\frac{1}{x}$ is convex. Therefore, by Jensen's inequality,
$$\frac{f(a+b)+f(b+c)+f(a+c)}{3}\ge f\left(\frac{(a+b)+(b+c)+(a+c)}{3}\right),$$
which implies the first inequality. Also by Jensen's inequality, we have
$$\frac{f(a)+f(b)}{2}\ge f\left(\frac{a+b}{2}\right).$$

Taking the cyclic sum of this inequality over a, b, and c, we get the second inequality.

Problem 8

(a) Prove that $\mathbb{N}$ can be written as the union of three disjoint sets such that any $m, n\in\mathbb{N}$ with $|m-n|=2, 5$ lie in different sets.

(b) Prove that $\mathbb{N}$ can be written as the union of four disjoint sets such that any $m, n\in\mathbb{N}$ with $|m-n|=2, 3, 5$ lie in different sets. Also show that this cannot be done with three sets.

Solution. **(a)** It is easy to check that the sets
$$\{3k+1\}_{k\in\mathbb{N}},\ \{3k+2\}_{k\in\mathbb{N}},\ \{3k\}_{k\in\mathbb{N}}$$
satisfy the condition.

(b) First we notice that the sets
$$\{4k+1\}_{k\in\mathbb{N}},\ \{4k+2\}_{k\in\mathbb{N}},\ \{4k+3\}_{k\in\mathbb{N}},\ \{4k\}_{k\in\mathbb{N}}$$
satisfy the condition.

We prove the second statement by contradiction. Suppose that there are sets A, B, C that satisfy the condition. We notice that the numbers 1, 3, 6 must be in the different sets. WLOG, we suppose that $1\in A$, $3\in B$, $6\in C$. Then $4\in B$. We also notice that $2, 5\notin B$ and 2 and 5 are in different sets. This leads to two cases: $\{1, 2\}\subset A$, $\{3, 4\}\subset B$,

$\{5,6\} \subset C$ or $\{1,5\} \subset A$, $\{3,4\} \subset B$, $\{2,6\} \subset C$. But for both cases, we can not put 7 in any of the three sets. This completes our proof.

Problem 9

A sequence of real numbers x_n is defined recursively as follows: x_0, x_1 are arbitrary positive real numbers, and

$$x_{n+2} = \frac{1 + x_{n+1}}{x_n}, \quad n = 0, 1, 2, \ldots.$$

Find x_{1998}.

Solution. We have

$$x_2 = \frac{1 + x_1}{x_0}, \quad x_3 = \frac{x_0 + x_1 + 1}{x_0 x_1}, \quad x_4 = \frac{1 + x_0}{x_1},$$

$x_5 = x_0$, and $x_6 = x_1$. Therefore, x_k periodically repeats every 5 terms and $x_{1998} = x_3 = \dfrac{x_0 + x_1 + 1}{x_0 x_1}$.

Problem 10

A triangle ABC has positive integer sides, $\angle A = 2\angle B$ and $\angle C > 90°$. Find the minimum length of the perimeter of ABC.

First Solution. This is USAMO 1991/1. The answer is 77. Let $BC = a$, $CA = b$, $AB = c$. We have $A = 2B$ and $C = 180° - 3B$. By the law of sines,

$$\frac{b}{\sin B} = \frac{a}{\sin A} = \frac{c}{\sin C}.$$

Since $\sin A = \sin 2B = 2\sin B \cos B$ and $\sin C = \sin 3B = 3\sin B - 4\sin^3 B$, we have

$$a = 2b\cos B, \quad c = b(3 - 4\sin^2 B) = b(4\cos^2 B - 1)$$

and hence $a^2 = b(b + c)$. Since we are looking for a triangle of smallest perimeter, we may assume that $\gcd(a, b, c) = 1$. In fact, $\gcd(b, c) = 1$, since any common factor of b and c would be a factor of a as well. We notice that since a perfect square a^2 is being expressed as the product of two relatively prime integers b and c, it must be the case that both b and $b{+}c$ are perfect squares. Thus, for some integers m and m, with $\gcd(m, m) = 1$, we have $b = m^2$, $b + c = n^2$, $a = mn$, $2\cos B = n/m = a/b$. Since $C > 90°$, we have $0 < B < 30°$ and

$$\sqrt{3} < 2\cos B = \frac{n}{m} < 2.$$

It is easy to check that $(m, n) = (4, 7)$ is the smallest pair that generates a triangle $(a, b, c) = (28, 16, 33)$ that meets all the conditions.

Second Solution. We use the same notations as those in the first solution. Let the angle bisector of $\angle CAB$ meet BC at D. Since $\angle BAD = \angle ABD$, we let $AD = BD = x$. We have $\angle ACD = \angle B$, $\angle ACB = \angle ACD$, so triangles ABC and DAC are similar. We have

$$\frac{x}{c} = \frac{b}{a} = \frac{a - x}{b}$$

which leads to $ax = bc$, $b^2 = a^2 - ax \Longrightarrow a^2 = b(b + c)$, and the rest is the same.

1.9 Japan

Problem 1

Let $p \geq 3$ be a prime, and let p points $A_0, \ldots, A_{p-1}$ lie on a circle in that order. Above the point $A_{1+\cdots+k-1}$ we write the number k for $k = 1, \ldots, p$ (so 1 is written above A_0). How many points have at least one number written above them?

Solution. Let congruences all be taken modulo p. The k-th triangular number is $t_k = k(k+1)/2$. Let $0 \leq s_k \leq p-1$ and $s_k \equiv t_k$. Therefore $k+1$ is written over the point A_{s_k} for $k = 0, 1 \ldots, p-1$. Since

$$\begin{aligned} t_n \equiv t_k &\Longleftrightarrow n(n+1) = k(k+1) \\ &\Longleftrightarrow n^2 + n - k^2 - k \equiv (n-k)(n+k+1) \equiv 0 \\ &\Longleftrightarrow n \equiv k \end{aligned}$$

or $n \equiv p - 1 - k$. Therefore

$$A_{s_0} = A_{s_{p-1}}, A_{s_1} = A_{s_{p-2}}, \ldots, A_{s_{(p-1)/2}} = A_{s_{(p-1)/2}}$$

and the answer is

$$\frac{p-1}{2} + 1 = \frac{p+1}{2}.$$

Problem 2

A country has 1998 airports connected by some direct flights. For any three airports, some two are not connected by a direct flight. What is the maximum number of direct flights that can be offered?

Solution. The answer is $999^2 = 998001$. It follows from the following general result.

Lemma 12. *A triangle-free graph G with $2n$ vertices has at most n^2 edges. If G has n^2 edges, then G is isomorphic to $K_{n,n}$ the complete bipartite graph with n vertices in each class.*

Proof. The proof is by induction. The result is evident for $n = 1$, so we take $n > 1$ and assume that the result is true for a triangle-free graph with $2(n-1)$ vertices. Given that G is triangle free and has $2n$ vertices, let v_1 and v_2 be two adjacent vertices, and let d_1 and d_2 be their respective degrees. Since G is triangle-free, $d_1 + d_2 \leq 2n$. Deleting v_1 and v_2 and their incident edges, we have a triangle-free graph G' with $2n-2$ vertices. Thus G has at most

$$(n-1)^2 + d_1 + d_2 - 1 \leq n^2$$

edges. Moreover, if G has n^2 edges then G' is isomorphic to $K_{n-1,n-1}$ and $d_1 + d_2 = 2n$. let U_1 and U_2 be the vertex classes of G'. Since there is no triangle, $d_1 \leq n$ and $d_2 \leq n$, so $d_=d_2 = n$, and we may assume that v_i is completely joined to U_i for $i = 1, 2$. Therefore G is isomorphic to $K_{n,n}$.

Problem 3

Let $P_1, \ldots, P_n$ be the sequence of vertices of a closed polygons whose sides may properly intersect each other at points other than the vertices. The external angle at P_i is defined as $180°$ minus the angle of rotation about P_i required to bring the ray P_iP_{i-1} onto the ray P_iP_{i+1}, taken in the range $(0°, 360°)$. (Here $P_0 = P_n$ and $P_1 = P_{n+1}$). Prove that if the sum of the external angles is a multiple of $720°$, then the number of self-intersections is odd.

Solution. For a polygon P, let $E(P)$ denote the sum of its external angles. For three points A, B, C let (ABC) denote the external angle at B – $180°$ minus the angle of rotation about B required to bring the ray BA onto the ray BC, taken in the range $(0°, 360°)$. We first establish the following lemma.

Lemma 13. *If there are k self-intersections in polygon P, then* $E(P) \equiv (k+1) \times 360° \pmod{720°}$.

Proof. The proof is by strong induction on k. For $k = 0$, the sum of the external angles is $\pm 360°$. Now take $n = k > 1$ and suppose that the result is true for all $n \leq k-1$. Suppose that there are k self-intersections in P. At least a pair of segments must intersect. WLOG, for some $i < j$, P_iP_{i+1} meets P_jP_{j+1} at P. Here $P_i = P_{n+i}$. Let polygon P_1 be $P_1, \cdots P_iPP_{j+1} \cdots P_n$ and polygon $P_2 = PP_{i+1} \cdots P_j$. Thus the total self-intersections in P_1 and P_2 is $k-1$. If P_1 has l self-intersections, then P_2 has $k-l-1$ self-intersections. From induction hypothesis, modulo $720°$,

$$E(P_1) \equiv (l+1) \times 360° \quad \text{and} \quad E(P_2) \equiv (k-l-1+1) \times 360°.$$

Noticing that $(P_iPP_{j+1}) = -(P_{i+1}PP_j)$, we have

$$\begin{aligned} E(P) &= E(P_1) + E(P_2) - (P_iPP_{j+1}) - (P_{i+1}PP_j) \\ &\equiv (k+1) \times 360^\circ \pmod{720^\circ}. \end{aligned}$$

This completes the induction step and therefore the proof of the lemma. □

So if the sum of the external angles is a multiple of 720°, then k the number of self-intersections is odd.

Problem 4

Let $c_{n,m}$ be the number of permutations of $\{1, \ldots, n\}$ which can be written as the product of m transpositions of the form $(i, i+1)$ for some $i = 1, \ldots, n-1$ but not of $m-1$ such transpositions. Prove that for all $n \in \mathbb{N}$,

$$\sum_{m=0}^{\infty} c_{n,m} t^m = \prod_{i=1}^{n} (1 + t + \cdots + t^{i-1}).$$

Solution. We proceed inductively on n.

For $n = 1$, there is 1 permutation that can be reached by 0 transpositions and both sides of the equation is equal to 1.

Assume that the equation is true for $n = k$. For $n = k+1$. Let $p_{k+1,m}$ be a permutation of $\{1, \ldots, k+1\}$ which can be written as the product of m transpositions but not of $m-1$ transpositions and $k+1$ is in the $k+1-h$-th position. Then it takes at least h transpositions to get $k+1$ to the $k+1-h$-th position. Therefore $p_{k+1,m} - k + 1$ is a permutation of $\{1, \ldots, k\}$ which can be written as the product of $m-h$ transpositions but not of $m-h-1$ transpositions. Clearly, this correspondence is a bijection. So we have $c_{k+1,m} = c_{k,m} + c_{k,m-1} + \cdots + c_{k,m-k}$. Therefore

$$\begin{aligned} \prod_{i=1}^{k+1} (1 + t + \cdots + t^{i-1}) &= \left(\sum_{m=0}^{\infty} c_{k,m} t^m \right) (1 + t + \cdots + t^k) \\ &= \sum_{m=0}^{\infty} (c_{k,m} + c_{k,m-1} + \cdots + c_{k,m-k}) t^m \\ &= \sum_{m=0}^{\infty} c_{k+1,m} t^m \end{aligned}$$

and the induction is complete.

Problem 5

On each of 12 points around a circle we place a disk with one white side and one black side. We may perform the following move: select a black disk, and reverse its two neighbors. Find all initial configurations from which some sequence of such moves leads to the position where all disks but one are white.

Solution. Label the discs as $d_1, d_2 \ldots, d_{12}$ in a clockwise order where disc d_k is the same as disc d_{k+12}. Let b denote the number of black discs. We consider the following cases.

(a) b is even. After each move, there will still be an even number of black discs. Therefore, it is impossible to do if there are an even number of black discs.

(b) n is odd and black discs and white discs are split in two groups. WLOG, let $d_1, \ldots, d_{2i+1}$ be black and $d_{2i+2}, \ldots, d_{12}$ be white. Then reversing the discs around $d_{2i}, d_{2i-2}, \ldots, d_2$ in that order will lead the position where $d_2, \ldots, d_{2i}$ are black and the others are white. We can then renumber the discs so d_2 is the first disc and repeat the process if necessary and we will end up with one black disc.

(c) n is odd and black discs and white discs are mixed. WLOG, let d_6 be a black disc and $d_i, \ldots, d_j$ be the block of black discs around d_6, i.e., $i \leq 6 \leq j$ and both d_{i-1} and d_{j+1} are white. Let d_k be the black disc closest to d_j. Therefore, $d_{j+1}, \ldots, d_{k-1}$ are all white. Flip the discs around d_k then d_{k-1} now is black. Thus the distance between the block and the next black is getting smaller. Repeat the process if necessary. We will end up with case (b).

From the above, we see that it is possible to obtain the position where all disks but one are white, as long as there are an odd number of black discs to start with.

1.10 Korea

Problem 1

Find all pairwise relatively prime positive integers l, m, n such that

$$(l+m+n)\left(\frac{1}{l}+\frac{1}{m}+\frac{1}{n}\right)$$

is an integer.

Solution. Answer: $(1,1,1)$, $(1,1,2)$, $(1,2,3)$ and all permutations. Bringing the second term to a common denominator, we get $lmn \mid (l+m+n)(lm+mn+nl)$ and, in particular, $l \mid (l+m+n)(lm+mn+nl) \Rightarrow l \mid (m+n)mn \Rightarrow l \mid m+n$; likewise $m \mid n+l$ and $n \mid l+m$. Now assume WLOG that $l \geq m \geq n$. Then $(m+n)/l = 1$ or 2; if the latter we have $l = m = n = 1$. In the case $m+n = l$, we get $m \geq l/2 \Rightarrow l/m \leq 2$ and $(n+l)/m \leq 3$. Moreover $l \geq m$ so this ratio cannot be 1; hence $n+l = 2m$ giving the $(1,2,3)$ solution or $n+l = 3m$ giving the $(1,1,2)$ solution.

Problem 2

Let D, E, F be points on the sides BC, CA, AB, respectively, of triangle ABC. Let P, Q, R be the second intersections of AD, BE, CF, respectively, with the circumcircle of ABC. Show that

$$\frac{AD}{PD}+\frac{BE}{QE}+\frac{CF}{RF} \geq 9$$

and determine when equality occurs.

Solution. Fix ABC and note that $\frac{AD}{PD} = \frac{d(A,BC)}{d(P,BC)}$, which has a constant numerator and so is minimized when the denominator is maximized, which occurs when P is the midpoint of the arc BC; and analogously for Q and R. Hence it suffices to prove the result when rays AD, BE, CF are angle bisectors. We have $\angle PBD = \angle BAC/2 = \angle PAB$ and so triangles PBD, PAB are similar and

$$\frac{PA}{PD} = \frac{PA}{PB}\cdot\frac{PB}{PD} = \left(\frac{PA}{PB}\right)^2 = \left(\frac{AB}{BD}\right)^2.$$

But using the angle bisector theorem, $AB/BD = (b+c)/a$ and likewise $BC/CE = (c+a)/b$, $CA/AF = (a+b)/c$; now either expanding, regrouping, and using AM-GM or, more elegantly, using RMS-AM and

AM-GM, as shown below, gives

$$\sum \frac{PA}{PD} = \sum \left(\frac{AB}{BD}\right)^2 \geq \frac{1}{3}\left(\sum \frac{AB}{BD}\right)^2$$
$$= \frac{1}{3}\left(\sum \frac{b+c}{a}\right)^2 \geq 12$$

and subtracting 3 from both sides gives our result. Equality requires that AD, BE, CF be angle bisectors and (because of the AM-GM step) that ABC be equilateral.

Problem 3

For a natural number n, let $\phi(n)$ denote the number of natural numbers less than or equal to n and relatively prime to n, and let $\psi(n)$ be the number of prime factors of n. Show that if $\phi(n)$ divides $n-1$ and $\psi(n) \leq 3$, then n is prime.

Solution. Note that for prime p, if $p^2 \mid n$ then $p \mid \phi(n)$ but $p \nmid n-1$, contradiction. So we need only show that $n \neq pq, n \neq pqr$ for primes $p < q < r$.

First assume $n = pq$, so $(p-1)(q-1) \mid pq-1$. Note that $q \geq 3$ implies that the left side is even, so the right is too and p, q are odd. But if $p = 3$, $q = 5$ then

$$(pq-1)/(p-1)(q-1) < 2;$$

the left side is decreasing in each variable and always > 1 so it cannot be an integer, contradiction.

Now let $n = pqr$. As before p, q, r are odd; if $p = 3$, $q = 7$, and $r = 11$ then

$$(pqr-1)/(p-1)(q-1)(r-1) < 2$$

and again the left side is decreasing and > 1; this eliminates all cases except where $p = 3, q = 5$. Then for $r = 7$ we have

$$(pqr-1)/(p-1)(q-1)(r-1) < 3$$

so the only integer value ever attainable is 2. Note that $(15r-1)/8(r-1) = 2$ gives $r = 15$ which is not a prime and we have eliminated all cases.

Problem 4

For positive real numbers a, b, c with $a+b+c=abc$, show that

$$\frac{1}{\sqrt{1+a^2}}+\frac{1}{\sqrt{1+b^2}}+\frac{1}{\sqrt{1+c^2}}\le\frac{3}{2},$$

and determine when equality occurs.

Solution. Let $\alpha=\tan^{-1}a$, $\beta=\tan^{-1}b$, $\gamma=\tan^{-1}c$ and the given conditions become $\alpha,\beta,\gamma>0$, $\alpha+\beta+\gamma=\pi$. What we want is now $\cos\alpha+\cos\beta+\cos\gamma\le 3/2$, which follows from Jensen's inequality.

Problem 5

Let I be the incenter of triangle ABC, O_1 a circle passing through B and tangent to the line CI at I, and O_2 a circle passing through C and tangent to the line BI at I. Show that O_1, O_2 and the circumcircle of ABC pass through a single point.

Solution. We direct angles mod π. Let O_1 and O_2 intersect at I and J. Tangency implies

$$\angle BJI=\angle IJC=\angle BIC=\frac{\pi-\angle CAB}{2}.$$

So $\angle BJC=\pi-\angle CAB=\angle BAC$ and A, B, C, J are concyclic.

Problem 6

Let F_n be the set of all bijective functions from $\{1,\ldots,n\}$ to $\{1,\ldots,n\}$ satisfying

(a) $f(k)\le k+1$ $(k=1,\ldots,n)$;

(b) $f(k)\ne k$ $(k=2,\ldots,n)$.

Find the probability that $f(1)\ne 1$ for f randomly chosen from F_n.

Solution. Answer: $\bar{F}_{n-1}/\bar{F}_n$, where $\bar{F}_n$ is the nth Fibonacci number (to avoid confusion). By induction we show that F_n has $\bar{F}_n$ elements and $\bar{F}_{n-1}$ of them satisfy $f(1)=2$. For $n=1$ this is apparent; now let $n\ge 2$. If $f\in F_n$ and $f(1)=2$ then we can define a function $g\in F_{n-1}$ by $g(k)=1$ if $f(k+1)=1$ and $g(k)=f(k+1)-1$ otherwise; conversely, each $g\in F_{n-1}$ corresponds to a unique f with $f(1)=2$. So the number of such f is the cardinality of F_{n-1} which is $\bar{F}_{n-1}$ by induction. On the other hand, the set of $f\in F_n$ with $f(1)=1$ is in 1-1 correspondence with the $g \in F_{n-1}$ satisfying $g(1) = 2$, as is seen by letting $g(k) = f(k+1) - 1$. Thus the number of such f is $\bar{F}_{n-2}$ using the induction hypothesis, and the total number of elements of F_n is $\bar{F}_{n-1} + \bar{F}_{n-2} = \bar{F}_n$ completing the induction.

1.11 Poland

Problem 1

Find all integers (a, b, c, x, y, z) such that

$$a + b + c = xyz$$
$$x + y + z = abc$$

and $a \geq b \geq c \geq 1$, $x \geq y \geq z \geq 1$.

First Solution. First we claim that at least one of bc and yz has its value less than 3. If $bc = 3$, then $b = 3, c = 1, a + b + c < 3a = abc$; if $bc > 3$, then $abc > 3a \geq a+b+c$. Thus for $bc \geq 3$, we have $abc > a+b+c$ and

$$3x \geq x + y + z = abc > a + b + c = xyz \Longrightarrow 3 > yz.$$

This proves our claim. WLOG, suppose that $yz = 1$ or 2.

If $yz = 1$, then $y = z = 1$. We have

$$abc = x + y + z = x + 2 = xyz + 2 = a + b + c + 2.$$

If $c \geq 2$, then $bc \geq 4$ and $4a \leq abc = a + b + c + 2 \leq 4a$; thus $a = b = c = 2$. We obtain the solutions $(2, 2, 2, 6, 1, 1)$ and $(6, 1, 1, 2, 2, 2)$. If $c = 1$, then $ab = a + b + 3$. If $b \geq 3$, then $3a \leq ab = a + b + 3 \leq 3a \Longrightarrow a = b = 3$. We obtain the solutions $(3, 3, 1, 7, 1, 1)$ and $(7, 1, 1, 3, 3, 1)$. If $b = 2$, we have $a = 5$ and obtain the solutions $(5, 2, 1, 8, 1, 1)$ and $(8, 1, 1, 5, 2, 1)$. If $b = 1$, we have $a = a + 4$, which is impossible.

If $yz = 2$, then $y = 2, z = 1$. We have

$$2abc = 2(x + y + z) = 2x + 6 = xyz + 6 = a + b + c + 6 \leq 3a + 6.$$

If $c \geq 2$, then $8a \leq 2abc \leq 3a + 6 \Longrightarrow 5a < 6$, which contradicts the fact that $a \geq c$. Thus $c = 1$, and $2ab = a + b + 7$. If $b \geq 3$, $6a \leq 2ab = a + b + 7 \Longrightarrow a \leq b/5 + 7/5$, which contradicts the fact that $a \geq b$. If $b = 2$, then $4a = 2ab = a + 9$ and $a = 3$. We obtain the solution $(3, 2, 1, 3, 2, 1)$. If $b = 1$, we have $a = 8$, repeating the solution $(8, 1, 1, 5, 2, 1)$.

Second Solution. Let

$$A = (ab - 1)(c - 1), B = (a - 1)(b - 1)$$
$$X = (xy - 1)(z - 1), Y = (x - 1)(y - 1).$$

Thus A, B, X, Y are nonnegative integers such that

$$A + B + X + Y = 4.$$

Clearly, neither of c and z can be greater than 2; that would force either A or Y to be greater than 4, and contradict the fact that $A+B+X+Y=4$.

If $c = 2$, we have $a, b \geq 2$ and $A \geq 3, B \geq 1$. Thus $A = 3, B = 1, X = Y = 0$. This yields the solution $(2,2,2,6,1,1)$. Similarly, if $z = 2$, we have $(6,1,1,2,2,2)$ as a solution.

Now we suppose that $c = z = 1$. We have $A = X = 0$ and $B+Y = 4$. WLOG, suppose that $Y \leq B$, (i.e., $Y = 0, 1, 2$.)

If $Y = 0$, we have $B = (a-1)(b-1) = 4$. This leads to the solutions $(5,2,1,8,1,1)$ and $(3,3,1,7,1,1)$. By symmetry, we also have the solutions $(8,1,1,5,2,1)$ and $(7,1,1,3,3,1)$.

If $Y = 1$, then $x = y = 2$ and $B = (a-1)(b-1) = 3 \Longrightarrow a = 4, b = 2$, but $a + b + c = 7 \neq xyz$.

If $Y = 2$, then $(x-1)(y-1) = (a-1)(b-1) = 2 \Longrightarrow a = x = 3$, $b = y = 2$. We obtain $(3,2,1,3,2,1)$ as our last solution.

Problem 2

The Fibonacci sequence F_n is given by

$$F_0 = F_1 = 1, F_{n+2} = F_{n+1} + F_n \quad (n = 0, 1, \ldots).$$

Determine all pairs (k, m) of integers with $m > k \geq 0$, for which the sequence x_n defined by $x_0 = F_k/F_m$ and

$$x_{n+1} = \frac{2x_n - 1}{1 - x_n} \text{ if } x_n \neq 1, \qquad x_{n+1} = 1 \text{ if } x_n = 1$$

contains the number 1.

Solution. We claim that $(k, m) = (2l, 2l+1)$ for all nonnegative integral l. We have

$$x_{n+1} = \frac{2x_n - 1}{1 - x_n} \Longrightarrow x_n = \frac{x_{n+1} + 1}{x_{n+1} + 2}$$

We proceed with backwards induction. We know that for some j, $x_j = 1 = F_0/F_1$. Suppose that $x_{p+1} = F_{2i}/F_{2i+1}$. Then,

$$x_p = \frac{F_{2i}/F_{2i+1} + 1}{F_{2i}/F_{2i+1} + 2} = \frac{F_{2i+2}}{F_{2i+3}}.$$

which implies that $x_0 = F_{2l}/F_{2l+1}$ and $(k, m) = (2l, 2l+1)$, for some integer $l \geq 0$. For any nonnegative integer l, it is easy to check that $x_0 = F_{2l}/F_{2l+1}$ generates a sequence that meets the condition.

Problem 3

The convex pentagon $ABCDE$ is the base of the pyramid $ABCDES$. A plane intersects the edges SA, SB, SC, SD, SE in points A', B', C', D', E', respectively, which differ from the vertices of the pyramid. Prove that the intersection points of the diagonals of the quadrilaterals $ABB'A'$, $BCC'B'$, $CDD'C'$, $DEE'D'$, $EAA'E'$ are coplanar.

Solution. Let A_1, B_1, C_1, D_1, E_1 be the intersection points of the diagonals of the quadrilaterals $ABB'A'$, $BCC'B'$, $CDD'C'$, $DEE'D'$, $EAA'E'$, respectively. Let ω_1 and ω_2 denote plane $ABCDE$ and plane $A'B'C'D'E'$ respectively.

If ω_1 is parallel to ω_2, then it is clear that A_1, B_1, C_1, D_1, E_1 are in a plane parallel to both ω_1 and ω_2.

Now suppose that ω_1 and ω_2 are not parallel. Let ω_1 meet ω_2 at line ℓ. Let ω denote the plane determined by A_1 and ℓ. Line AC is on the planes ω_1 and $SA'ACC'$, and line $A'C'$ is on the planes ω_2 and $SA'ACC'$. Thus, the lines AC, $A'C'$, ℓ are concurrent; let K be their intersection. Notice that A_1 and B_1 are on the planes $AB'C$ and $A'BC'$. So, the planes meet at the line A_1B_1. But K is on both AC and $A'C'$, so K is on both planes $AB'C$ and $A'BC'$. It is clear that plane $AB'C$ is not parallel to $A'BC'$, and so they intersect in a line. Thus, A_1, B_1, K are collinear, implying that $B_1 \in \omega$. Similarly, we can show that C_1 lies on the plane determined by B_1 and ℓ, the same plane as ω. Thus A_1, B_1, C_1, D_1, E_1 are on ω.

Problem 4

Prove that the sequence $\{a_n\}$ defined by $a_1 = 1$ and

$$a_n = a_{n-1} + a_{\lfloor n/2 \rfloor} \quad n = 2, 3, 4, \ldots$$

contains infinitely many integers divisible by 7.

Solution. We prove the statement by contradition. Suppose that there are only finitely many integer terms in the sequence that are divisible by 7 and that a_k is the last such element in the sequence. Then observe that

$$a_{2k-1} \equiv a_{2k} \equiv a_{2k+1} \equiv a \not\equiv 0 \pmod{7}$$

Then, the seven elements beginning with a_{4k-3} will all have different residues modulo 7. Actually, for $n = 0, 1, \ldots, 6$, $a_{4k-3+n} \equiv a_{4k-3} + na \pmod 7$. This yields a contradiction. So there are infinitely many terms divisible by 7.

Problem 5

Points D, E lie on side AB of the triangle ABC and satisfy

$$\frac{AD}{DB} \cdot \frac{AE}{EB} = \left(\frac{AC}{CB}\right)^2.$$

Prove that $\angle ACD = \angle BCE$.

First Solution. Applying the law of sines translates the problem into

$$\frac{AD}{DB} \cdot \frac{AE}{EB} = \left(\frac{\sin B}{\sin A}\right)^2.$$

Let $\theta = \angle ACD, \phi = \angle ECB, \alpha = \angle DCE$. Again, application of the law of sines yields that $CD/\sin A = AD/\sin\theta$ and $DE/\sin B = BE/\sin\phi$. Dividing these two equations,

$$\frac{CD \sin B}{CE \sin A} = \frac{AD \sin\phi}{BE \sin\theta}$$

$$\iff \frac{CD \sin B}{CE \sin A} = \frac{BD \sin^2 B \sin\phi}{AE \sin^2 A \sin\theta}$$

$$\iff \frac{CD}{CE} = \frac{BD \sin B \sin\phi}{AE \sin A \sin\theta}$$

$$\iff \frac{CD \cdot AE \sin A}{CE \cdot DB \sin B} = \frac{\sin\phi}{\sin\theta}$$

But

$$\frac{CD}{BD} = \frac{\sin B}{\sin(\phi + \alpha)} \quad \text{and} \quad \frac{AE}{CE} = \frac{\sin(\theta + \alpha)}{\sin A}.$$

Thus,

$$\frac{\sin A \sin B \sin(\theta + \alpha)}{\sin B \sin(\phi + \alpha) \sin A} = \frac{\sin\phi}{\sin\theta}$$

$$\iff \sin(\theta + \alpha)\sin\theta = \sin\phi \sin(\phi + \alpha)$$

$$\iff \cos(2\theta + \alpha) - \cos\alpha = \cos(2\phi + \alpha) - \cos\alpha$$

$$\iff \cos(2\theta + \alpha) = \cos(2\phi + \alpha)$$

Since $(2\theta+\alpha)+(2\phi+\alpha)=2\angle C<360^\circ$, $2\theta+\alpha=2\phi+\alpha$ and hence $\angle ACD=\theta=\phi=\angle BCE$.

Second Solution. Extend AC and let F and G be points on the ray AC such that $BF \parallel CE$ and $BG \parallel CD$. We have $\angle CGB=\angle ACD$ and $\angle CBF=\angle BCE$. We claim that $\angle CBF=\angle CGB$, which implies that $\angle ACD=\angle BCE$. We rewrite our condition as

$$\frac{AC^2\cdot BD\cdot BE}{AD\cdot AE}=BC^2.$$

We notice that

$$\begin{aligned}
&\frac{CF}{AC}=\frac{BE}{AE},\ \frac{CG}{AC}=\frac{BD}{AD},\\
\Longrightarrow\ &\frac{AC^2\cdot BD\cdot BE}{AD\cdot AE}=CF\cdot CG\\
\Longrightarrow\ &BC^2=CF\cdot CG\\
\Longleftrightarrow\ &\frac{BC}{CF}=\frac{CG}{BC}.
\end{aligned}$$

We also notice that $\angle BCF=\angle GCB$; thus triangle CFB is similar to triangle CBG by SAS, implying that $\angle CBF=\angle CGB$ as desired.

Problem 6

Consider unit squares in the plane whose vertices have integer coordinates. Let S be the chessboard which contains all unit squares lying entirely inside the circle $x^2+y^2\le 1998^2$. In each square of S we write $+1$. A move consists of reversing the signs of a row, column or diagonal of S. Can we end up with exactly one square containing -1?

Solution. The answer is no. By symmetry, each row in S has an even number of squares. Before a particular row move, suppose that the sum of all the numbers on the row is $2n-2k$, i.e., there are k squares numbered -1 and $2n-k$ squares numbered 1 on the row. After the move, the row sum is $2k-2n$ and the sum of the all the numbers changes by a multiple of 4. An analogous fact is true about the columns and diagonals. Therefore, the sum of all the numbers remains constant modulo 4. To have exactly one square contain -1 would require our total sum to change by 2 modulo 4; this violates our invariant.

1.12 Romania

Problem 1

Let A_n denote the set of words of length n formed from the letters a, b, c which do not contain two consecutive letters both equal to a or both equal to b. Let B_n denote the set of words of length n formed from the letters a, b, c which do not contain three consecutive letters which are pairwise distinct. Prove that $|B_{n+1}| = 3|A_n|$ for all $n \geq 1$.

First Solution. Let X_n, Y_n, Z_n be subsets of A_n that contain words that end up with letters a, b, c, respectively. For positive integer n, we have the following relations.

$$\begin{aligned}
|A_n| &= |X_n| + |Y_n| + |Z_n|; \\
|X_{n+1}| &= |Y_n| + |Z_n|; \\
|Y_{n+1}| &= |X_n| + |Z_n|; \\
|Z_{n+1}| &= |X_n| + |Y_n| + |Z_n|; \\
|Z_{n+1}| &= |A_n|,
\end{aligned}$$

which implies that

$$\begin{aligned}
|A_{n+2}| &= |X_{n+2}| + |Y_{n+2}| + |Z_{n+2}| \\
&= 2|X_{n+1}| + 2|Y_{n+1}| + 3|z_{n+1}| = 2|A_{n+1}| + |A_n|,
\end{aligned}$$

with $A_1 = 3, A_2 = 7$.

Let S_n be subsets of B_n that contain words that end up with letters aa, bb, cc. Let $T_n = B_n - S_n$, i.e., the T_n contains those words in S_n that end up with ab, ac, ba, bc, ca, cb. For positive integer n, we have the following relations.

$$\begin{aligned}
|B_n| &= |S_n| + |T_n|; \\
|B_{n+1}| &= 3|S_n| + 2|T_n|; \\
|S_{n+1}| &= |B_n|,
\end{aligned}$$

which implies that $|B_{n+2}| = |S_{n+2}| + |T_{n+2}| = 2|B_{n+1}| + |B_n|$, with $|B_2| = 3^2 = 9 = 3|A_1|$, $|B_3| = 3^3 - 3! = 21 = 3|A_2|$. Thus $|B_{n+1}| = 3|A_n|$.

Second Solution. Replace a, b, and c by 1, 2, and 0 respectively, and words of length n by ordered n-tuples. Let S_n be the set of all ordered n-tuples in which each term is 0, 1, or 2. Consider the function $\Delta : S_{n+1} \to S_n$ defined as follows:

$$\Delta(x_1, \ldots, x_{n+1}) = (y_1, \ldots, y_n)$$

where $y_i \equiv x_{i+1} - x_i \pmod 3$. Observe that x_{i-1}, x_i, x_{i+1} are pairwise distinct if and only if $x_i - x_{i-1} = x_{i+1} - x_i \in \{1, 2\}$, so $(x_1, \ldots, x_{n+1})$ contains no three consecutive distinct elements if and only if $\Delta(x_1, \ldots, x_{n+1})$ contains no two consecutive 1's or 2's. Therefore $\Delta(B_{n+1}) = A_n$. Since $|\Delta^{-1}(\{y\})| = 3$ for all $y \in S_{n-1}$, we have $|B_{n+1}| = 3|A_n|$.

Problem 2

The volume of a parallelepiped is 216 cubic centimeters and its surface area is 216 square centimeters. Prove that the parallelepiped is a cube.

Solution. First, suppose the parallelepiped is rectangular with edge lengths x, y, z centimeters. Then

$$2(xy + yz + zx) \geq 6(xyz)^{2/3} = 6(216)^{2/3} = 216$$

by AM-GM, so the surface area is at least 216 cm^2, with equality if and only if $x = y = z$, i.e., the parallelepiped is a cube. Now consider a non-rectangular parallelepiped $ABCDA'B'C'D'$ in which the "top" face $A'B'C'D'$ is not directly above the "bottom" face $ABCD$. Then moving $A'B'C'D'$ above $ABCD$ leaves the volume fixed and decreases the surface area. Repeating this for each pair of faces gives a rectangular parallelepiped with strictly smaller surface area and volume 216 cm^3. By the previous part, this rectangular parallelepiped has surface area at least 216 cm^2, so the original parallelepiped had surface area greater than 216 cm^2. Thus if a parallelepiped has volume 216 cm^3 and surface area 216 cm^2, it must be a cube.

Problem 3

Let $m \geq 2$ be an integer. Find the smallest integer $n > m$ such that for any partition of the set $\{m, m+1, \ldots, n\}$ into two subsets, at least one subset contains three numbers a, b, c (not necessarily different) such that $a^b = c$.

Solution. The smallest such integer is $m^{m^{m+2}}$. First, suppose A and B are a partition of $\{m, m+1, \ldots, m^{m^{m+2}}\}$ into two subsets such that neither subset contains three numbers a, b, c such that $a^b = c$. Assume WLOG $m \in A$; then $m^m \in B$, $(m^m)^{(m^m)} = m^{m^{m+1}} \in A$, and $\left(m^{m^{m+1}}\right)^m = m^{m^{m+2}} \in B$. Now consider the element m^{m+1}. If it is in set A, then we can take $a = m$, $b = m^{m+1}$, and $c = m^{m^{m+1}}$, a contradiction. If it is in set B, take $a = m^m$, $b = m^{m+1}$, and $c = m^{m^{m+2}}$, again a contradiction. So $n = m^{m^{m+2}}$ satisfies the desired property.

Now we exhibit a partition of $\{m, m+1, \ldots, m^{m^{m+2}} - 1\}$ into two subsets such that neither subset contains three numbers a, b, c such that $a^b = c$. Take

$$A = \{m, m+1, \ldots, m^m - 1\} \cup \{m^{m^{m+1}}, \ldots, m^{m^{m+2}} - 1\}$$

and $B = \{m^m, \ldots, m^{m^{m+1}} - 1\}$. Then for any a, $b \in B$, $a^b \geq (m^m)^{(m^m)} = m^{m^{m+1}}$ so $a^b \notin B$. Also, there are no a, $b \in A$ such that $a^b \in A$ because of the inequalities

$$m^m > m^m - 1, \quad (m^m - 1)^{(m^m - 1)} < (m^m)^{(m^m)} = m^{m^{m+1}},$$
$$\left(m^{m^{m+1}}\right)^m > m^{m^{m+2}} - 1.$$

Thus $m^{m^{m+2}}$ is the smallest such n.

Problem 4

Consider a finite set of segments in the plane whose lengths have sum less than $\sqrt{2}$. Prove that there exists an infinite unit square grid whose lines do not meet any of the segments.

Solution. For a given line ℓ, let $\pi(\ell)$ denote the length of the projection of the given segments onto ℓ. Let ℓ_1 and ℓ_2 be two perpendicular lines, and consider $\pi(\ell_1) + \pi(\ell_2)$. A segment of length x contributes at most $x\sqrt{2}$ to this sum, so $\pi(\ell_1) + \pi(\ell_2) < 2$ as the total length of the given segments is less than $\sqrt{2}$.

Now we will show $\pi(L_1) = \pi(L_2)$ for some two perpendicular lines L_1 and L_2. Pick any two perpendicular lines ℓ_1 and ℓ_2; if $\pi(\ell_1) = \pi(\ell_2)$ we are done, so assume WLOG $\pi(\ell_1) > \pi(\ell_2)$. Now rotate ℓ_1 and ℓ_2 by $90°$ continuously to ℓ_1' and ℓ_2'; then $\ell_1' \parallel \ell_2$ and $\ell_2' \parallel \ell_1$, so $\pi(\ell_1') < \pi(\ell_2')$. Thus by continuity there exist some perpendicular lines L_1 and L_2 such that $\pi(L_1) = \pi(L_2)$. In particular, since $\pi(L_1) + \pi(L_2) < 2$, we have $\pi(L_1) < 1$ and $\pi(L_2) < 1$.

Finally, since the projection of the given segments onto L_1 has length less than 1, we can find a set S of lines parallel to L_2 spaced 1 unit apart which do not meet any of the given segments. To do so, we toss in the plane a set of lines S_1 parallel to L_2 spaced one unit apart. Since we are given finitely many segments, we can find a subset $S_2 \subset S_1$ of n the consecutive lines $\ell_1, \dots \ell_n$ (in that order from left to right) such that all the given segments are in between ℓ_1 and ℓ_n. Let L_1 meet $\ell_1, \dots, \ell_n$ at $P_1, \dots, P_n$, respectively. For $i = 2, \dots, n-1$, we translate intervals $(P_i, P_{i+1}]$ to coincide with $(P_1, P_2]$. Thus we obtain a set $T \subset [P_1, P_2]$ by translating the projections of all the segments onto L_1 to $[P_1, P_2]$. Set T is a union of segments. Since the total length of T is less than or equal to $\pi(L_1) < 1$ which is the length of $[P_1, P_2]$, we can find a point P such that $P \in [P_1, P_2]$ and $P \notin T$. We translate S_1 by $\overrightarrow{P_1P}$ to obtain desired set S.

Similarly, we can find a set of lines parallel to L_1 spaced 1 unit apart which do not meet any of the given segments. The union of these lines gives the desired grid.

Problem 5

We are given an isosceles triangle ABC with $BC = a$ and $AB = AC = b$. The points M, N vary along AC, AB, respectively, so as always to satisfy the condition

$$a^2 \cdot AM \cdot AN = b^2 \cdot BN \cdot CM.$$

The lines BM and CN meet at P. Find the locus of P.

Solution. Let γ be the circle tangent to AB and AC at B and C; then the locus of P is the arc of γ inside triangle ABC. Suppose $M \in AC$ and $N \in AB$ are such that $a^2 \cdot AM \cdot AN = b^2 \cdot BN \cdot CM$. Construct $M' \in AB$, $N' \in AC$ such that $MM' \parallel NN' \parallel BC$; then $MM' = (a/b)AM$, $NN' = (a/b)AN$, so $MM' \cdot NN' = BN \cdot CM$ and $BN/NN' = MM'/CM$. Also

$$\angle BNN' = \pi - \angle ABC = \pi - \angle ACB = \angle M'MC,$$

so triangles BNN' and $M'MC$ are similar. Therefore

$$\angle PCB = \angle NCB = \angle NN'B = \angle MCM' = \angle MBM' = \angle ABP$$

and P lies on the arc of γ inside triangle ABC. Now, let P be any point on this arc, and let $M = AC \cap BP$ and $N = AB \cap CP$. Then

$\angle ABM = \angle BCN$ and $\angle ACN = \angle CBM$, so

$$\frac{AM \cdot AN}{BN \cdot CM} = \frac{AM}{MC} \cdot \frac{AN}{NB} = \frac{b \sin \angle ABM}{a \sin \angle MBC} \cdot \frac{b \sin \angle ACN}{a \sin \angle NCB} = \frac{b^2}{a^2}$$

and P is in the desired locus.

Problem 6

The vertices of a convex pentagon are all lattice points. Prove that the area of the pentagon is at least $5/2$.

Note. This is Putnam 1990/A-3. We introduce three methods. The first two methods use very basic knowledge and the the third method uses Pick's theorem. Let $ABCDE$ be a convex lattice pentagon with area K. We have the following lemmas.

Lemma 14. *At least one of the midpoints of the sides and the diagonals of $ABCDE$ is a lattice point.*

Proof. Some two of A, B, C, D, E have both coordinates congruent modulo 2, so their midpoint is a lattice point M inside or on pentagon $ABCDE$. □

Lemma 15. *The area of a non-degenerate lattice triangle can be written in the form of $m/2$, where m is an positive integer.*

Proof. The triangle can be inscribed into a rectangle with all its sides parallel to the coordinate axes. The area of the triangle is equal to the area of the rectangle minus the total area of three (or two) right triangles. Our results follows from the fact that lengths of the legs of each right triangle and the sides of the rectangle are all integers. □

First Solution. We consider the following two cases.

(a) M is inside $ABCDE$, then $ABCDE$ can be cut into 5 lattice triangles. By the second lemma, each triangle has area at least $1/2$ and $K \geq 5/2$.

(b) M is on $ABCDE$, say M is on AB. Since $ABCDE$ is convex, AB can not parallel to both CD and DE. WLOG, assume that $AB \not\parallel CD$. Then $[ACD]$, $[MCD]$, $[BCD]$ must form either a strictly increasing or a strictly decreasing sequence. From the second lemma,

$$\max\{[ACD], [MCD], [BCD]\} \geq \frac{3}{2}.$$

Then $ABCDE$ can be cut into that triangle with the largest area and two other lattice triangles. From the second lemma, $K \geq 3/2 + 1/2 + 1/2 = 5/2$.

Second Solution. Our desired result follows from (a) in the first proof and the following lemma.

Lemma 16. *There is at least 1 lattice point inside lattice pentagon $ABCDE$.*

Proof. From the first lemma, we know that there is a lattice point M on or inside $ABCDE$. If M is inside we are done. Suppose that M is on $ABCDE$. WLOG, say M is on AB. Let B' be the lattice point on AB that is closest to A. We clip triangle $BB'C$ away. Then $AB'CDE$ is a convex lattice pentagon and $AB'CDE$ is within the boundary of $ABCDE$. Apply the first lemma to $AB'CDE$, if lattice point M' is inside $AB'CDE$ then we are done. If M' is on the sides of $AB'CDE$, it will not be on AB' by the definition of B'. We can repeat the above procedure to $AB'CDE$. With at most 5 steps, we can obtain a lattice pentagon (within the boundary of $ABCDE$) that either has a lattice inside it or has a lattice on its sides but not on the sides of $ABCDE$. For either case, we find a lattice point that is inside $ABCDE$. □

Third Solution. Let I the number of lattice points in its interior, and P the number of lattice points on its perimeter. Then from Pick's theorem, $K = I + P/2 - 1$. We have $I + P \geq 6$ and $K \geq 6/2 - 1 = 2$ for any convex lattice pentagon. If M is inside the pentagon, then $I \geq 1$ and $P \geq 5$ so $K \geq 1 + 5/2 - 1 = 5/2$. Otherwise, M lies on the boundary of the pentagon; suppose WLOG M is the midpoint of AB. Then $AMCDE$ is a convex lattice pentagon, so its area is at least 2; thus $K = [AMCDE] + [MBC] \geq 2 + 1/2 = 5/2$.

Problem 7

Find all positive integers (x, n) such that $x^n + 2^n + 1$ is a divisor of $x^{n+1} + 2^{n+1} + 1$.

Solution. The solutions are $(x, n) = (4, 1)$ and $(11, 1)$. If $n = 1$, we need $x + 3 = x + 2 + 1 \mid x^2 + 4 + 1 = x^2 + 5 = (x+3)(x-3) + 14$, so $x + 3$ divides 14 and $x = 4$ or 11. Suppose $n \geq 2$. For $x \in \{1, 2, 3\}$ we have

$$1 + 2^n + 1 < 1 + 2^{n+1} + 1 < 2(1 + 2^n + 1),$$

$$2^n + 2^n + 1 < 2^{n+1} + 2^{n+1} + 1 < 2(2^n + 2^n + 1),$$
$$2(3^n + 2^n + 1) < 3^{n+1} + 2^{n+1} + 1 < 3(3^n + 2^n + 1),$$

so $x^n + 2^n + 1$ does not divide $x^{n+1} + 2^{n+1} + 1$. For $x \geq 4$, $x^n = x^n/2 + x^n/2 \geq 2^{2n}/2 + x^2/2$, so

$$\begin{aligned}(2^n + 1)x &\leq ((2^n + 1)^2 + x^2)/2 \\ &= (2^{2n} + 2^{n+1} + 1 + x^2)/2 < 2^{n+1} + x^n + 2^n + 2.\end{aligned}$$

Therefore

$$\begin{aligned}(x-1)(x^n + 2^n + 1) &= x^{n+1} + 2^n x + x - x^n - 2^n - 1 \\ &< x^{n+1} + 2^{n+1} + 1 < x(x^n + 2^n + 1);\end{aligned}$$

again $x^n + 2^n + 1$ does not divide $x^{n+1} + 2^{n+1} + 1$. So the only solutions are $(4, 1)$ and $(11, 1)$.

Problem 8

Let $n \geq 2$ be an integer. Show that there exists a subset A of $\{1, 2, \ldots, n\}$ having at most $2\lfloor\sqrt{n}\rfloor + 1$ elements such that

$$\{\, |x - y| : x, y \in A,\ x \neq y\,\} = \{1, 2, \ldots, n - 1\}.$$

Solution. Let $k = \lfloor\sqrt{n}\rfloor$, and let m be the largest integer such that $m(k+1) < n$. Then $m \leq k$, since $(k+1)^2 > n$. It is easy to check that

$$\{1, 2, \ldots, k+1, 2(k+1), 3(k+1), \ldots, m(k+1), n\}$$

has the desired properties.

Problem 9

Show that for any positive integer n, the polynomial

$$f(x) = (x^2 + x)^{2^n} + 1$$

cannot be written as the product of two nonconstant polynomials with integer coefficients.

Solution. Note that $f(x) = g(h(x))$, where $h(x) = x^2 + x$ and $g(y) = y^{2^n} + 1$. Since

$$g(y+1) = (y+1)^{2^n} + 1 = y^{2^n} + \left(\sum_{k=1}^{2^n - 1} \binom{2^n}{k} y^k\right) + 2,$$

and $\binom{2^n}{k}$ is even for $1 \le k \le 2^n - 1$, g is irreducible by Eisenstein's criterion. Now let p be a nonconstant factor of f, and let r be a root of p. Then $g(h(r)) = f(r) = 0$, so $s := h(r)$ is a root of g. Since $s = r^2 + r \in \mathbb{Q}(r)$, we have $\mathbb{Q}(s) \subset \mathbb{Q}(r)$, so

$$\deg p \ge \deg(\mathbb{Q}(r)/\mathbb{Q}) \ge \deg(\mathbb{Q}(s)/\mathbb{Q}) = \deg g = 2^n.$$

Thus every factor of f has degree at least 2^n. Therefore, if f is reducible, we can write $f(x) = g(x)h(x)$ where g and h have degree 2^n.

Next, observe

$$\begin{aligned} f(x) &\equiv (x^2 + x)^{2^n} + 1 \\ &\equiv x^{2^{n+1}} + x^{2^n} + 1 \equiv (x^2 + x + 1)^{2^n} \pmod 2; \end{aligned}$$

since $x^2 + x + 1$ is irreducible in $\mathbb{Z}_2[x]$, by unique factorization we must have

$$g(x) \equiv h(x) \equiv (x^2 + x + 1)^{2^{n-1}} \equiv x^{2^n} + x^{2^{n-1}} + 1 \pmod 2.$$

Thus, if we write

$$\begin{aligned} g(x) &= a_{2^n} x^{2^n} + \cdots + a_0, \\ h(x) &= b_{2^n} x^{2^n} + \cdots + b_0, \end{aligned}$$

then $a_{2^n}, a_{2^{n-1}}, a_0, b_{2^n}, b_{2^{n-1}}, b_0$ are odd and all the other coefficients are even. Since f is monic, we may assume without loss of generality that $a_{2^n} = b_{2^n} = 1$; also, $a_0 b_0 = f(0) = 1$, but $a_0 > 0$, $b_0 > 0$ as f has no real roots, so $a_0 = b_0 = 1$. Therefore,

$$\begin{aligned} &([x^{2^n + 2^{n-1}}] + [x^{2^{n-1}}])(g(x)h(x)) \\ &\qquad \equiv \left(\sum_{i=2^{n-1}}^{2^n} a_i b_{2^n + 2^{n-1} - i} \right) + \left(\sum_{i=0}^{2^{n-1}} a_i b_{2^{n-1} - i} \right) \\ &\qquad \equiv a_{2^n} b_{2^{n-1}} + a_{2^{n-1}} b_{2^n} + a_0 b_{2^{n-1}} + a_{2^{n-1}} b_0 \\ &\qquad \equiv 2(a_{2^{n-1}} + b_{2^{n-1}}) \\ &\qquad \equiv 0 \pmod 4 \end{aligned}$$

as $a_{2^{n-1}} + b_{2^{n-1}}$ is even. But

$$([x^{2^n + 2^{n-1}}] + [x^{2^{n-1}}])(f(x)) = \binom{2^n}{2^{n-1}} = 2\binom{2^n - 1}{2^{n-1} - 1},$$

and $\binom{2^n-1}{2^{n-1}-1}$ is odd by Lucas's Theorem, so

$$\left([x^{2^n+2^{n-1}}]+[x^{2^{n-1}}]\right)(f(x)) \equiv 2 \pmod 4,$$

a contradiction. Hence f is irreducible.

Problem 10

Let $n \geq 3$ be a prime number and $a_1 < a_2 < \cdots < a_n$ be integers. Prove that $a_1, a_2, \ldots, a_n$ is an arithmetic progression if and only if there exists a partition of $\{0, 1, 2, \ldots\}$ into classes $A_1, A_2, \ldots, A_n$ such that

$$a_1 + A_1 = a_2 + A_2 = \cdots = a_n + A_n.$$

(Note: $x + S$ denotes the set $\{x + y : y \in S\}$.)

Solution. First, suppose $a_1, a_2, \ldots, a_n$ is an arithmetic progression $a_i = k + id$. Then the sets

$$A_i = \{0, 1, \ldots, d-1\} + \{0, nd, 2nd, \ldots\} + (n-i)d$$

partition $\{0, 1, 2, \ldots\}$, and

$$a_i + A_i = \{0, 1, \ldots, d-1\} + \{0, nd, 2nd, \ldots\} + nd + k$$

for all i, so we are done.

Now we will prove the converse. Call a pair of sets of nonnegative integers (A, B) an (a, b)-cover if $|A| = a$, $|B| = b$, and $A + B = \{0, 1, \ldots, ab-1\}$. Observe that if (A, B) is an (a, b)-cover, then $0 \in A \cap B$, and as $|A + B| = ab = |A| \cdot |B|$, every element of $|A + B| = \{0, 1, \ldots, ab-1\}$ can be written as $x+y$ with $x \in A$ and $y \in B$ in exactly one way; in particular $|A \cap B| = \{0\}$.

Lemma 17. *Let (A, B) be an (a, b)-cover and let*

$$m = \min(B \setminus \{0\}).$$

Then m divides a and there exists an $(a/m, b)$-cover (A', B') such that $A = mA' + \{0, 1, \ldots, m-1\}$, $B = mB'$.

Proof. We claim that for all $k = im + j$ with $i \geq 0$ and $0 \leq j < m$,

$$k \in A \Leftrightarrow im \in A \quad \text{and} \quad k \in B \Rightarrow j = 0.$$

The proof is by induction on k. Clearly $\{0, 1, \ldots, m-1\} \subset A$ and $\{0, m\} \subset B$, so the statement holds for all $0 \leq k \leq m$. Now let $k > m$

and suppose the claim holds for all smaller values of k. If $j = 0$, there is nothing to show, so assume $0 < j < m$.

First, suppose $k \in A$ but $im \notin A$. By the inductive hypothesis, $k - 1 \notin A$, so $k - 1 = x + y$ for $x \in A$, $y \in B$, $y > 0$. Then $y \geq m$, so $x \leq k - 1 - m$. If $x + 1 \in A$, then $(x + 1) + y = k = k + 0$, a contradiction, so $x + 1 \notin A$. So by the inductive hypothesis $x + 1$ must be a multiple of m. But then $y = k - 1 - x \equiv k \not\equiv 0 \pmod{m}$, and $y \in B$, $y \leq k - 1$, contradicting the inductive hypothesis. So $k \in A \Rightarrow im \in A$.

Next, suppose $im \in A$ but $k \notin A$. Then $k - 1 \in A$ by the inductive hypothesis, and $k = x + y$ for some $x \in A$, $y \in B$, $x < k$. Then $x - 1 \notin A$, as otherwise $(x - 1) + y = k - 1 = (k - 1) + 0$, a contradiction. Thus by the inductive hypothesis x is a multiple of m. So $y = k - x$ is not a multiple of m; as $y \in B$, the only possibility is $y = k$, $x = 0$, so $k \in B$. But then $m - j \in A$, $k \in B$, $im \in A$, $m \in B$, and $(m - j) + k = im + m$, a contradiction. So $im \in A \Rightarrow k \in A$.

Finally, suppose $k \in B$. We can write $im = x + y$ for some $x \in A$, $y \in B$. If $y = 0$, then $im = x \in A$, $m \in B$, $m - j \in A$, $k \in B$, and $im + m = (m - j) + k$, a contradiction; hence $y > 0$. Then $y \geq m$ and $y \leq im < k$, so y is a multiple of m by the inductive hypothesis. Thus $x \leq (i - 1)m$, x is a multiple of m and $x + j < im < k$, so by the inductive hypothesis $x + j \in A$. Therefore $(x + j) + y = k = 0 + k$, a contradiction. So $k \notin B$ if $0 < j < m$.

Thus the claim holds for all $k \geq 0$. It follows that

$$A = mA' + \{0, 1, \ldots, m - 1\} \text{ and } B = mB'$$

for some sets A', B' of positive integers. Clearly $|A'| = |A|/m = a/m$, $|B'| = |B| = b$, and

$$A' + B' = \{0, 1, \ldots, ab/m - 1\}$$

as

$$\{0, 1, \ldots, ab - 1\} = A + B = m(A' + B') + \{0, 1, \ldots, m - 1\}.$$

Hence (A', B') is an $(a/m, b)$-cover.

Clearly this lemma also holds with the roles of A and B reversed. □

Lemma 18. *Let (A, B) be an (a, b)-cover such that a is prime. Then the elements of A form an arithmetic sequence.*

Proof. We prove this by induction on b. If $b = 1$, then $B = \{0\}$ so $A = A + B = \{0, 1, \ldots, a - 1\}$ and the statement is obvious. So suppose

$b > 1$ and assume the lemma holds for all smaller b. Then $ab \geq 2$, so $1 \in A + B$; thus $1 \in A$ or $1 \in B$.

(1) $1 \in A$. Then $1 \notin B$, so $m := \min(B \setminus \{0\}) > 1$. Then, by the first lemma, m divides a and there exists an $(a/m, b)$-cover (A', B') such that

$$A = mA' + \{0, 1, \ldots, m-1\} \text{ and } B = mB'.$$

Since $m > 1$ divides a, a prime, $m = a$ and $|A'| = a/m = 1$ so $A' = \{0\}$. Thus

$$A = m\{0\} + \{0, 1, \ldots, m-1\} = \{0, 1, \ldots, m-1\},$$

and the elements of A form an arithmetic sequence.

(2) $1 \in B$. Then $1 \notin A$, so $m := \min(A \setminus \{0\}) > 1$. By the first lemma, m divides b and there exists an $(a, b/m)$-cover (A', B') such that

$$A = mA' \text{ and } B = mB' + \{0, 1, \ldots, m-1\}.$$

Then a is prime and $b/m < b$, so by the inductive hypothesis the elements of A' form an arithmetic sequence. Since $A = mA'$, the elements of A also form an arithmetic sequence. □

Now suppose we have a partition $\{A_i\}$ of $\{0, 1, 2, \ldots\}$ such that $a_1 + A_1 = \cdots = a_n + A_n$. Let $c_i = a_n - a_i$, $C = \{c_1, \ldots, c_{n-1}, c_n\}$ and $S = a_i + A_i - a_n = A_i - c_i$. Note that $c_1 > \cdots > c_n = 0$ and $S \subset \{0, 1, 2, \ldots\}$. Define $C(x) = \sum_{i \in C} x^i$ and $S(x) = \sum_{i \in S} x^i$. Also define $s_i = [i \in S]$. Because the sets $A_i = S + c_i$ partition $\{0, 1, 2, \ldots\}$, we have $C(x)S(x) = \frac{1}{1-x}$ and $\sum_{j=1}^{n} s_{i-c_j} = 1$ for all $i \geq 0$. Therefore $s_i = 1 - s_{i-c_1} - \cdots - s_{i-c_{n-1}}$ so s_i is determined by the previous c_1 values s_{i-c_1}, s_{i-c_1+1}, $\ldots$, s_{i-1}. Since there are only finitely many possible sequences for c_1 consecutive values, there must exist some $m \geq 0$, $k \geq 0$ such that

$$s_m = s_{m+k},\ s_{m+1} = s_{m+k+1},\ \ldots,\ s_{m+c_1-1} = s_{m+k+c_1-1};$$

it follows that $s_{m+i} = s_{m+k+i}$ for all $i \geq 0$, so the sequence $\{s_i\}$ is eventually periodic with period k. Moreover,

$$s_{i+c_1-c_1} + s_{i+c_1-c_2} + \cdots + s_{i+c_1-c_n} = 1,$$

so

$$s_i = s_{i+c_1-c_1} = 1 - s_{i+c_1-c_2} - \cdots - s_{i+c_1-c_n}$$

for all $i \geq 0$; hence s_i can also be determined from the following c_1 terms $s_{i+1}, \ldots, s_{i+c_1-c_n} = s_{i+c_1}$. Thus the sequence $\{s_i\}$ is in fact periodic with period k, so $s_i = s_{i+k}$ for all $i \geq 0$. Therefore $S = D+\{0, k, 2k, \ldots\}$ for some set $D \subset \{0, 1, \ldots, k-1\}$. Let $D(x) = \sum_{i\in D} x^i$; then $S(x) = D(x)/(1-x^k)$ and

$$\frac{1}{1-x} = C(x)S(x) = \frac{C(x)D(x)}{1-x^k}$$

so $C(x)D(x) = 1+x+\cdots+x^{k-1}$. Therefore (C, D) is an $(n, k/n)$-cover. Since n is prime, it follows from the second lemma that the elements of C form an arithmetic sequence. Hence $(a_1, \ldots, a_n) = (a_n - c_1, \ldots, a_n - c_n)$ is an arithmetic progression.

Problem 11

Let n be a positive integer and let P_n denote the set of integer polynomials of the form $a_0 + a_1x + \cdots + a_nx^n$, where $|a_i| \leq 2$ for $i = 0, 1, \ldots, n$. Find, for each positive integer k, the number of elements of the set $A_n(k) = \{f(k) : f \in P_n\}$.

Solution. For $k = 1$, $A_n(k)$ contains $4n+5$ elements; for $2 \leq k \leq 4$, $A_n(k)$ contains $4(k^{n+1}-1)/(k-1)+1$ elements; for $k \geq 5$, $A_n(k)$ contains 5^{n+1} elements.

First, suppose $k \geq 5$. Let $f(x) = a_0 + a_1x + \cdots + a_nx^n$, $g(x) = b_0 + b_1x + \cdots + b_n$ be two distinct elements of P_n. Then there exists $0 \leq m \leq n$ such that $a_n = b_n$, $a_{n-1} = b_{n-1}$, $\ldots$, $a_{m+1} = b_{m+1}$, $a_m \neq b_m$. Therefore

$$|a_mk^m - b_mk^m| \geq k^m,$$

but

$$\begin{aligned} &|(a_0-b_0) + (a_1-b_1)k + \cdots + (a_{m-1}-b_{m-1})k^{m-1}| \\ &\leq 4 + 4k + \cdots + 4k^{m-1} = 4\frac{k^m-1}{k-1} \\ &\leq k^{m-1} < k^m, \end{aligned}$$

so $a_0 + a_1k + \cdots + a_mk^m \neq b_0 + b_1k + \cdots + b_mk^m$ and

$$f(k) = a_0 + a_1k + \cdots + a_nk^n \neq b_0 + b_1k + \cdots + b_nk^n = g(k).$$

Thus the numbers $f(k)$, $f \in P_n$ are all distinct, so

$$|A_n(k)| = |P_n| = 5^{n+1} \quad \text{for } k \geq 5.$$

Now, suppose $k \leq 4$. Let $[a, b]$ denote $\{ c \in \mathbb{Z} : a \leq c \leq b \}$. We will prove that $A_n(k) = [-M_n, M_n]$ where

$$M_n = 2 + 2k + \cdots + 2k^n$$

by induction on n. For $n = 0$, $A_n(k) = \{-2, -1, 0, 1, 2\}$ and $M_n = 2$. Now let $n \geq 1$ and assume $A_{n-1}(k) = [-M_{n-1}, M_{n-1}]$. Then

$$\begin{aligned} A_n(k) &= \{a_0 + a_1 k + \cdots + a_n k^n : |a_i| \leq 2 \text{ for all } i\} \\ &= A_{n-1}(k) + \{a_n k^n : |a_n| \leq 2\} \\ &= (A_{n-1}(k) - 2k^n) \cup (A_{n-1}(k) - k^n) \cup A_{n-1}(k) \\ &\qquad \cup (A_{n-1}(k) + k^n) \cup (A_{n-1}(k) + 2k^n) \\ &= [-M_{n-1} - 2k^n, M_{n-1} - 2k^n] \\ &\qquad \cup [-M_{n-1} - k^n, M_{n-1} - k^n] \\ &\qquad \cup \cdots \cup [-M_{n-1} + 2k^n, M_{n-1} + 2k^n]; \end{aligned}$$

since $2M_{n-1} \geq k^n$ (obvious for $k = 1$, and for $1 < k \leq 4$, $2M_{n-1} = 4(k^n - 1)/(k - 1) > k^n - 1$), these intervals overlap and $A_n(k) = [-M_{n-1} - 2k^n, M_{n-1} + 2k^n] = [-M_n, M_n]$. Thus by induction $A_n(k) = [-M_n, M_n]$ for all n, so

$$|A_n(k)| = 2M_n + 1 = \begin{cases} 4n + 5, & \text{if } k = 1; \\ 4\frac{k^n - 1}{k - 1} + 1, & \text{for } 2 \leq k \leq 4. \end{cases}$$

Problem 12

Find all functions $u : \mathbb{R} \to \mathbb{R}$ for which there exists a strictly monotonic function $f : \mathbb{R} \to \mathbb{R}$ such that

$$f(x + y) = f(x)u(y) + f(y) \quad \forall x, y \in \mathbb{R}.$$

Solution. The solutions are $u(x) = e^{kx}$, $k \in \mathbb{R}$. To see that these work, take $f(x) = x$ for $k = 0$. If $k \neq 0$ take $f(x) = e^{kx} - 1$; then

$$f(x + y) = e^{k(x+y)} - 1 = (e^{kx} - 1)e^{ky} + e^{ky} - 1 = f(x)u(y) + f(y)$$

for all x, $y \in \mathbb{R}$.

Now suppose $u : \mathbb{R} \to \mathbb{R}$, $f : \mathbb{R} \to \mathbb{R}$ are functions for which f is strictly monotonic and $f(x + y) = f(x)u(y) + f(y)$ for all x, $y \in \mathbb{R}$. We must show that u is of the form $u(x) = e^{kx}$ for some $k \in \mathbb{R}$. First, letting $y = 0$, we obtain $f(x) = f(x)u(0) + f(0)$ for all $x \in \mathbb{R}$; thus $u(0) \neq 1$

would imply $f(x) = f(0)/(1-u(0))$ for all x, so we must have $u(0) = 1$ and $f(0) = 0$. Then $f(x) \neq 0$ for all $x \neq 0$. Next, we have

$$f(x)u(y) + f(y) = f(x+y) = f(x) + f(y)u(x)$$

so

$$f(x)(u(y)-1) = f(y)(u(x)-1) \quad \text{for all } x, y \in \mathbb{R}.$$

Thus for any $x \neq 0$, $y \neq 0$, $(u(x)-1)/f(x) = (u(y)-1)/f(y)$, so there exists $C \in \mathbb{R}$ such that $(u(x)-1)/f(x) = C$ for all $x \neq 0$. So $u(x) = 1 + Cf(x)$ for $x \neq 0$; since $u(0) = 1$, $f(0) = 0$, this equation also holds for $x = 0$. If $C = 0$, then $u(x) = 1$ for all x and we are done. Otherwise, observe

$$\begin{aligned} u(x+y) &= 1 + Cf(x+y) \\ &= 1 + Cf(x)u(y) + f(y) \\ &= u(y) + Cf(x)u(y) \\ &= u(x)u(y) \end{aligned}$$

for all x, $y \in \mathbb{R}$. Thus $u(nx) = u(x)^n$ for all $n \in \mathbb{Z}$, $x \in \mathbb{R}$. Since $u(x) = 1 + Cf(x)$ for all x, u is strictly monotonic, and $u(-x) = 1/u(x)$ for all x, so $u(x) > 0$ for all x as $u(0) = 1$. Let $e^k = u(1)$; then $u(n) = e^{kn}$ for all $n \in \mathbb{N}$, and $u(p/q) = (u(p))^{1/q} = e^{k(p/q)}$ for all $p \in \mathbb{Z}$, $q \in \mathbb{N}$, so $u(x) = e^{kx}$ for all $x \in \mathbb{Q}$. Since u is monotonic and the rationals are dense in $\mathbb{R}$, we have $u(x) = e^{kx}$ for all $x \in \mathbb{R}$. Thus all solutions are of the form $u(x) = e^{kx}$, $k \in \mathbb{R}$.

Problem 13

On an $m \times n$ sheet of paper is drawn a grid dividing the sheet into unit squares. The two sides of length n are taped together to form a cylinder. Prove that it is possible to write a real number in each square, not all zero, so that each number is the sum of the numbers in the neighboring squares, if and only if there exist integers k, l such that $n+1$ does not divide k and

$$\cos\frac{2l\pi}{m} + \cos\frac{k\pi}{n+1} = \frac{1}{2}.$$

Solution. First, suppose there exist integers k, l such that the given conditions hold, and let

$$x_{ij} = \cos\frac{2il\pi}{m}\sin\frac{jk\pi}{n+1} \quad \text{for all } i,\, j \in \mathbb{Z}.$$

Then

$$\begin{aligned}
&x_{i-1,j} + x_{i+1,j} + x_{i,j-1} + x_{i,j+1} \\
&\qquad = \left(\cos \frac{2(i-1)l\pi}{m} + \cos \frac{2(i+1)l\pi}{m} \right) \sin \frac{jk\pi}{n+1} \\
&\qquad\quad + \cos \frac{2il\pi}{m} \left(\sin \frac{(j-1)k\pi}{n+1} + \sin \frac{(j+1)k\pi}{n+1} \right) \\
&\qquad = 2 \cos \frac{2l\pi}{m} \cos \frac{2il\pi}{m} \sin \frac{jk\pi}{n+1} \\
&\qquad\quad + 2 \cos \frac{j\pi}{n+1} \cos \frac{2il\pi}{m} \sin \frac{jk\pi}{n+1} \\
&\qquad = 2 \left(\cos \frac{2l\pi}{m} + \cos \frac{j\pi}{n+1} \right) \cos \frac{2il\pi}{m} \sin \frac{jk\pi}{n+1} \\
&\qquad = x_{ij}
\end{aligned}$$

for all i, $j \in \mathbb{Z}$; moreover, $x_{10} = \sin k\pi/(n+1) \neq 0$ since $n+1$ does not divide k, and $x_{ij} = x_{i+m,j}$ and $x_{0j} = x_{n+1,j} = 0$ for all i and j. Thus in the matrix

$$\begin{array}{cc|cccc|cc}
\cdots & x_{1,m-1} & x_{10} & x_{11} & \cdots & x_{1,m-1} & x_{10} & \cdots \\
\cdots & x_{2,m-1} & x_{20} & x_{21} & \cdots & x_{2,m-1} & x_{20} & \cdots \\
 & \vdots & \vdots & \vdots & & \vdots & \vdots & \\
\cdots & x_{n,m-1} & x_{n0} & x_{n1} & \cdots & x_{n,m-1} & x_{n0} & \cdots
\end{array}$$

each number is the sum of the numbers in the neighboring squares, and not all of the numbers are 0, so we are done.

Now, suppose we have an assignment of numbers (not all zero) to the squares of an $m \times n$ cylinder such that each number is the sum of the numbers in the neighboring squares. Let the m columns be represented by m vectors $x_0, \ldots, x_{m-1} \in \mathbb{R}^n$, numbered modulo m; then we have

$$x_i = x_{i-1} + x_{i+1} + Ax_i,$$

where A is the $n \times n$ matrix

$$A = \begin{bmatrix} 0 & 1 & & & \\ 1 & 0 & 1 & & \\ & 1 & 0 & & \\ & & & \ddots & 1 \\ & & & 1 & 0 \end{bmatrix}.$$

Since the x_i's are not all zero, we may assume without loss of generality that $x_0 \neq 0$. Now define $v_1, \ldots, v_n \in \mathbb{R}^n$ by

$$v_k = \left(\sin \frac{k\pi}{n+1}, \sin \frac{2k\pi}{n+1}, \ldots, \sin \frac{nk\pi}{n+1} \right);$$

observe that $v_k \neq 0$ (as $n+1$ does not divide k) and

$$\begin{aligned} Av_k &= \left(\sin \frac{(j-1)k\pi}{n+1} + \sin \frac{(j+1)k\pi}{n+1} \right)_{j=1}^{n} \\ &= \left(2\cos \frac{j\pi}{n+1} \sin \frac{jk\pi}{n+1} \right)_{j=1}^{n} \\ &= \left(2\cos \frac{k\pi}{n+1} \right) v_k, \end{aligned}$$

therefore v_k is an eigenvector of A with eigenvalue $r_k = 2\cos k\pi/(n+1)$. Since the eigenvalues r_k, $k = 1, \ldots, n$ are distinct, and $A^T = A$, for any distinct k, k', we have

$$r_k v_k^T v_{k'} = (Av_k)^T v_{k'} = v_k^T A^T v_{k'} = v_k^T A v_{k'} = r_{k'} v_k^T v_{k'},$$

so $v_k \cdot v_{k'} = v_k^T v_{k'} = 0$ and the vectors v_k are orthogonal. In particular, they are linearly independent, so they form a basis; thus as $x_1 \neq 0$, there exists $k \in \{1, \ldots, n\}$ such that $v_k \cdot x_0 \neq 0$. Define $y_i = v_k \cdot x_i$ for each i. Then $y_0 \neq 0$ and

$$y_i = v_k \cdot x_i = v_k \cdot x_{i-1} + v_k \cdot x_{i+1} + v_k \cdot Ax_i = y_{i-1} + y_{i+1} + r_k y_i,$$

so $y_{i-1} + (r_k - 1)y_i + y_{i+1} = 0$, or, writing $s = 1 - r_k$,

$$sy_i = y_{i-1} + y_{i+1} \quad \text{for all } i.$$

We claim $s = 2\cos 2l\pi/m$ for some integer l. First, note that this equation implies $|s| \cdot |y_i| = |y_{i-1} + y_{i+1}| \leq |y_{i-1}| + |y_{i+1}|$; summing this over $i = 0$, $\ldots, m-1$ gives $|s|S \leq 2S$, where $S = |y_0| + \cdots + |y_{m-1}| > 0$. Therefore $|s| \leq 2$, so we can write $s = 2\cos\alpha$ for some $\alpha \in \mathbb{R}$.

If $s = 2$, we can take $l = 0$; if $s = -2$, it is easy to see that m must be even, so we can take $l = m/2$. Otherwise, suppose $|s| < 2$, so $\cos\alpha \neq \pm 1$ and $\sin\alpha \neq 0$. Then the function $f : \mathbb{R} \setminus \cos^{-1}(\{0\}) \to \mathbb{R}$ defined by

$$\begin{aligned} f(\theta) &= \frac{\cos(\theta + \alpha)}{\cos\theta} = \frac{\cos\theta\cos\alpha - \sin\theta\sin\alpha}{\cos\theta} \\ &= \cos\alpha - \sin\alpha\tan\theta \end{aligned}$$

is surjective, so we can find $\theta \in \mathbb{R}$ such that $\cos\theta \neq 0$ and $f(\theta) = y_1/y_0$. Let $r = y_0/\cos\theta$; then $y_0 = r\cos\theta$ and $y_1 = r\cos(\theta+\alpha)$. For $i \geq 2$, it follows inductively that $y_i = \cos(\theta + i\alpha)$, since

$$\begin{aligned} y_{i+1} &= sy_i - y_{i-1} \\ &= 2\cos\alpha(r\cos(\theta+i\alpha)) - r\cos(\theta+(i-1)\alpha) \\ &= r(\cos(\theta+(i+1)\alpha) + \cos(\theta+(i-1)\alpha) - \cos(\theta+(i-1)\alpha)) \\ &= r\cos(\theta+(i+1)\alpha). \end{aligned}$$

However, $r\cos(\theta + tm\alpha) = y_{tm} = y_0 = r\cos\theta$ for all integers t, so $m\alpha$ must be a multiple of 2π and $\alpha = 2l\pi/m$ for some integer l. Thus we have shown that $s = 2\cos 2l\pi/m$ for some integer l.

Finally, since $2\cos 2l\pi/m = s = 1 - 2\cos k\pi/(n+1)$, we have found integers k and l such that $n+1$ does not divide k and

$$\cos\frac{2l\pi}{m} + \cos\frac{k\pi}{n+1} = \frac{1}{2}.$$

1.13 Russia

Problem 1

Do there exist n-digit numbers M and N such that all of the digits of M are even, all of the digits of N are odd, each digit from 0 to 9 occurs exactly once among M and N, and N divides M?

Solution. The answer is no. We proceed by indirect proof. Suppose that such M and N exist and let $a = M/N$. Then $M \equiv 0+2+4+6+8 \equiv 2 \pmod 9$ and $N \equiv 1+3+5+7+9 \equiv 7 \pmod 9$; they are both relatively prime to 9. Now $a \equiv M/N \equiv 8 \pmod 9$ and so $a \geq 8$. But $N \geq 13579$ so $M = aN \geq 8(13579) > 99999$, a contradiction.

Problem 2

In parallelogram $ABCD$, the points M and N are the midpoints of sides BC and CD, respectively. Can the lines AM and AN divide the angle BAD into three equal parts?

First Solution. The answer is no. We proceed by indirect proof. Let $AB = CD = s$ and $BC = DA = t$, and let G be the centroid of triangle ABC. We first show that $s = t$. $\angle ADN = \angle ABM$ since $ABCD$ is a parallelogram, and $\angle DAN = \angle BAM$ by assumption. Then triangles ADN and ABM are similar so

$$t/(s/2) = AD/DN = AB/BM = s/(t/2).$$

Thus, $s^2 = t^2$ and $s = t$, as claimed.

Now perform the translation that takes N to C and A to P, the midpoint of AB. Then AM and CP are medians of triangle ABC and meet at G. $\angle AGP = \angle MGC = \angle MAN = \angle PAG$, so $GP = AP = s/2$. Similarly, $GM = s/2$ so $GA = 2GM = s$. But then the sides of triangle APG are $s/2, s/2$, and s, violating the triangle inequality, a contradiction. Therefore our original assumption was false and AM and AN cannot trisect $\angle BAD$, as claimed.

Second Solution. Suppose that AM and AN trisect $\angle BAD$. Let line MN meet lines AB and AD at E and F respectively. Then $DN = NC$, $\angle FND = \angle CNM$ and $\angle FDN = \angle MCN$ ($AF \parallel CB$), which implies that triangles FDN and MCN are congruent (by ASA) and $FN = NM$. Now in $\triangle AFM$, AN is both the angle bisector and the median. Thus

$AF = AM$ and $AN \perp FM$. Similarly, we have $AM \perp NE$. In triangle AMN, there are two right angles; contradiction.

Problem 3

A deck contains 52 cards of 4 different suits. Vanya is told the number of cards in each suit. He picks a card from the deck, guesses its suit, and sets it aside; he repeats until the deck is exhausted. Show that if Vanya always guesses a suit having no fewer remaining cards than any other suit, he will guess correctly at least 13 times.

Solution. For the cards of each suit, label the cards $1, 2, \ldots$ in the reverse order from how they are picked from the deck (so that the 1 is the last card of each suit to be picked). By the Pigeonhole Principle, some suit has at least 13 cards. Then for each of $i = 1, 2, \ldots, 13$, there exists at least one card labeled i. For each i, look at the last card labeled i to be picked from the deck: say its suit is hearts. Then after the other cards numbered i are picked from the deck, Vanya will guess hearts until the i of hearts is picked, so he will guess its suit correctly.

Then for each i, Vanya guesses at least one suit correctly for at least 13 correct guesses in all.

Problem 4

In the plane are given $n \geq 9$ points. Given any 9 points, there exist two circles such that each of the 9 points lies on one of the circles. Prove there exist two circles such that each of the given points lies on one of the circles.

Solution. Consider some 9 points—by the pigeonhole principle, some 5 lie on a circle k: call them A, B, C, D, E. Now consider the points among our n points that are *not* on k.

If there are 3 or fewer, draw a circle through them; then all the n points lie on this circle or k, as desired.

If there are more than 3, choose any 3 (call them F, G, H) and draw a circle m through them. For each point I different from A through H, each of the nine points A through I lies on one of two circles. By the pigeonhole principle, some three of A, B, C, D, E lie on a circle; so, one circle must be k. And since F, G, H are not on k, the other must be m. Thus, I lies on k or m. Therefore, all of the points lie on k or m, as desired.

Problem 5

The side lengths of a triangle and the diameter of its incircle are 4 consecutive integers. Find all such triangles.

Solution. The only such triangle is a 3-4-5 triangle. Let r be the inradius of the triangle; a, b, c its side lengths; and s its semiperimeter. The smallest and largest of the four integers have the same sum as the middle two integers, so WLOG, we suppose that $2r + a = b + c$. Then $2r = b + c - a$ and $r = s - a$. Let I be the incenter of the triangle, and E and F be the points of tangency of the incircle with CA and AB. $IE = EA = AF = FI = s - a$. Thus $IEAF$ is a square and $\angle A = 90°$. So a must be the largest integer, $2r$ the smallest, and a, b, c are consecutive integers. The only such right triangle is a 3-4-5 triangle, in which case $2r = 2$ indeed works.

Problem 6

Two circles meet at P and Q. A line intersects segment PQ and meets the circles at the points A, B, C, D in that order. Prove that $\angle APB = \angle CQD$.

Solution. From the given information, C is between B and D. Also, since the line passes through segment PQ, convex quadrilaterals $APCQ$ and $BPDQ$ are cyclic. Then

$$\angle APB = \angle APQ - \angle BPQ = \angle ACQ - \angle BDQ = \angle DQC.$$

Problem 7

A 10-digit number is said to be interesting if its digits are all distinct and it is a multiple of 11111. How many interesting integers are there?

Solution. There are 3456 such integers.

Let $n = abcdefghij$ be a 10-digit interesting number. The digits of n must be $0, 1, \ldots, 9$, so modulo 9,

$$n \equiv a + b + c + d + e + f + g + h + i + j \equiv 0 + 1 + 2 + \cdots + 9 \equiv 0.$$

Since $\gcd(9, 11111) = 1$, $99999|n$. Let $x = abcde$ and $y = fghij$ be two 5-digit numbers. We have $n = 10^5x + y$. Thus

$$0 \equiv n \equiv 10^5 x + y \equiv x + y \pmod{99999}.$$

But $0 < x + y < 2(99999)$, so n is interesting if and only if $x + y = 99999$—that is, if $a + f = \cdots = e + j = 9$.

There are $5! = 120$ ways to distribute the pairs $(0,9)$, $(1,8)$, …, $(4,5)$ among $(a,f), (b,g), \ldots, (e,j)$, and for each pair we can swap the order of the digits: for example, (b,g) could equal $(0,9)$ or $(9,0)$. This gives $2^5 = 32$ more choices for a total of $(32)(120)$ numbers. However, one-tenth of these numbers have $a = 0$, which is not allowed. So, there are $(9/10)(32)(120) = 3456$ interesting numbers, as claimed.

Problem 8

We have a 102×102 sheet of graph paper and a connected figure of unknown shape consisting of 101 squares. What is the smallest number of copies of the figure which can be cut out of the square?

Solution. The answer is 4. We must show that (i) there exists a shape that we can get at most four copies of and (ii) we can get at least four copies of any shape.

(i) Consider a perfect "cross": a row of 51 squares and a column of 51 squares intersecting at each other's centers. We show we can cut at most four copies of it from the paper; suppose we can get five for sake of contradiction. Each cross's center lies in the middle 52×52 region of the graph paper, or else the cross would run off the paper. Then by the Pigeonhole Principle, two of the centers of these five shapes lie in the same 26×26 corner quadrant of this region, which is impossible.

(ii) One can prove by induction on n that any n-omino can be contained in a $k \times (n + 1 - k)$ rectangle of squares for some k between 1 and n. Then any 101-square shape can be fitted into a $k \times (102 - k)$ rectangle of squares for some k between 1 and 101. We can tile the border of the 102×102 square with four such rectangles: two in the upper-left and lower-right corners, and two (rotated 90°) in the upper-right and lower-left corners. From there we can cut out four shapes, as desired.

Problem 9

Let $f(x) = x^2 + ax + b\cos x$. Find all values of a, b for which the equations $f(x) = 0$ and $f(f(x)) = 0$ have the same (nonempty) set of real roots.

Solution. The answer is

$$0 \le a < 4 \text{ and } b = 0.$$

Let r be a root of $f(x)$. Then $b = f(0) = f(f(r)) = 0$, and thus $f(x) = x(x+a)$ and $r = 0, -a$. So $f(f(x)) = f(x)(f(x)+a) = x(x+a)(x^2+ax+a)$. We seek a such that x^2+ax+a has no real roots besides 0 and $-a$. If either 0 or $-a$ is a root of x^2+ax+a, we must have $a=0$ and $f(f(x))$ indeed has no other roots. Otherwise, the discriminant a^2-4a must be negative so $0<a<4$. Then $0 \le a < 4$, as claimed, in which case the two equations have the same set of roots: $x = 0$ and $-a$.

Problem 10

In acute triangle ABC, the circle S passes through the circumcenter O and vertices B, C. Let OK be a diameter of S, and let D, E be the second intersections of S with lines AB, AC, respectively. Show that $ADKE$ is a parallelogram.

Solution. All angles are directed modulo π. Note that OK is the perpendicular bisector of BC. We have

$$\angle KEC = \angle BDK = \frac{\angle COB}{2} = \angle BAC.$$

so $DK \parallel AC, EK \parallel AB$ and $ADKE$ is a parallelogram, as desired.

Problem 11

Show that from any finite set of points in the plane, one can remove one point so that the remaining set can be divided into two subsets, each with smaller diameter than the original set. (The diameter of a finite set of points is the maximum distance between two points in the set.)

Solution. Let q be the diameter of the set. Consider the points as vertices on a graph and draw an edge between points distance q apart.

Lemma 19. *There is no 4-cycle.*

Proof. Suppose by way of contradiction that $AB = BC = CD = DA = q$. Then $ABCD$ is a rhombus and one of its angles—say, $\angle ABC$—is at least 90°. Then applying the law of cosines on triangle ABC, we have $AC \ge q\sqrt{2}$, a contradiction. □

Lemma 20. *If there is a cycle $[M]$, then every edge shares a vertex with it.*

Proof. For any point X, define O_X as the circle centered at X with radius q. Also, call a point a *cycle-point* if it is in $[M]$.

For each cycle-point P, draw O_P; let R be the common interior/boundary region of these circles. R must contain all the points in the given set since no two points are more than a distance q apart from each other, and it is convex since it is the intersection of convex regions. Also, its boundary is made of minor arcs of the O_P and its vertices are points in the cycle.

Now assume by way of contradiction that $AB = q$ but that neither A nor B is in $[M]$. We look at two cases: (i) one of A or B is on the boundary of R, and (ii) neither is on the boundary of R.

(i) Assume without loss of generality that A is on the boundary of R. Then it is on the minor arc of a circle centered at some cycle-point C, bounded by some two cycle-points D and E. Let O_A intersect O_C at S and T, with S inside O_D but not in O_E, and T inside O_E but not in O_D.

Since $AB = q$, B lies on O_A; since $BC \leq q$, B does not lie outside O_C. Furthermore, since B is not a cycle-point, it cannot equal C.

Then B lies on either arc SC or CT of circle O_A. In the first case, it lies outside O_E so that $BE > q$, a contradiction; in the latter, it lies outside O_D so that $BD > q$, a contradiction.

(ii) Now suppose that AB lies in R's interior. Extend ray AB to meet R's boundary at P. If P is a vertex, then $PA > q$, a contradiction. Otherwise, mark off point Q on ray PA so that $PQ = q$; since Q is on segment PA and R is convex, Q is in R as well. Then PQ is a segment of length q contained by R, and P is on the boundary of R. However, this is impossible from above.

Then no matter where AB is, we have a contradiction. So, either A or B is a vertex, as claimed. □

Lemma 21. *There is at most one cycle.*

Proof. Assume otherwise by way of contradiction, and let $[P]$ be a cycle with minimal length. Any other cycle $[Q]$ must have at least one vertex V not in $[P]$. This vertex V is adjacent to some vertices W and X; from the second lemma, W and X are in $[P]$.

If two or more vertices separate W and X in cycle $[P]$, then replacing them in $[P]$ by V forms a smaller cycle, a contradiction.

If only one vertex Y separates W and X in cycle $[P]$, then $WYXV$ is a 4-cycle, which is impossible by the first lemma.

If no vertices separate them, then VWX is a 3-cycle. Since $[P]$ is a minimal cycle, it too must be a 3-cycle, with W and X both adjacent to some vertex Y. Then $WYXV$ is a 4-cycle, again impossible by lemma 1.

So, our original assumption was false and there is at most one cycle, as claimed. □

Now we prove our main result. If there is a cycle, remove any point in it. Otherwise, remove an arbitrary point. Since there is at most one cycle originally, there are no more cycles.

Each separate connected subgraph of the new graph must be a tree since there are no cycles left. In each subgraph, color an arbitrary point V red. Then from V to any other vertex W in the subgraph, there is a unique path. If the path has an odd number of edges, color W blue; otherwise, color W red. Then no two adjacent vertices have the same color.

Separate the points by color; then since no two adjacent points have the same color, no two points in the same set are at a distance q from each other, as desired.

Problem 12

In 1999 memory locations of a computer are stored the numbers $1, 2, \ldots, 2^{1998}$. Two programmers take turns subtracting 1 from each of 5 different memory locations. If any location ever acquires a negative number, the computer breaks and the guilty programmer pays for the repairs. Which programmer can ensure himself financial security, and how?

Solution. Name the first programmer Al and the second programmer Bob. Al can bankrupt Bob by first subtracting 1 from the locations with 1, 2^{1995}, 2^{1996}, 2^{1997}, and 2^{1998} on his first turn; then on subsequent turns subtracting 1 from the five locations that Bob just subtracted from.

On each turn, at least 1 is subtracted from $1, 2, \ldots, 2^{1994}$ so their total sum $2^{1995} - 1$ decreases by at least 1 on each turn. Then after at most 2^{1995} turns this sum would be negative so the computer would break.

But after at most 2^{1995} turns, the numbers that were originally 2^{1995}, 2^{1996}, 2^{1997}, and 2^{1998} would remain nonnegative. So, Al can never break the computer by subtracting 1 from these four locations; and if Bob subtracts 1 from any of these locations, Al can subtract 1 from the same locations on the next turn without breaking the computer.

As for the other locations, Al's first move leaves even numbers in the other 1995 locations. Whenever Bob subtracts 1 from any of these locations, he leaves odd numbers. If he leaves -1, he breaks the computer; otherwise, he leaves positive odd numbers that Al can safely subtract 1 from on his next turn, again leaving even numbers.

Therefore, Al never breaks the computer using this method; and he can ensure his financial security, as claimed.

Problem 13

Two matching decks have 36 cards each; one is shuffled and put on top of the second. For each card of the top deck, we count the number of cards between it and the corresponding card of the second deck. What is the sum of these numbers?

Solution. The answer is 1296. Label the pairs of matching cards $1, 2, \ldots, 36$. Let a_i be the location (from the bottom of the deck) of the top card labeled i, and b_i be the location of the bottom card labeled i. We wish to find $\sum_{i=1}^{36}(a_i - b_i)$ which is equal to

$$\sum_{i=1}^{36} a_i - \sum_{i=1}^{36} b_i = (37 + \ldots + 72) - (1 + \ldots + 36) = 36^2 = 1296.$$

Problem 14

A circle S centered at O meets another circle S' at A and B. Let C be a point on the arc of S contained in S'. Let E, D be the second intersections of S' with lines AC, BC, respectively. Show that $DE \perp OC$.

Solution. Let line OC meet DE at F. Then from cyclic quadrilaterals, we have

$$\begin{aligned}\angle FEC + \angle ECF &= \angle CBA + \angle ACO \\ &= \frac{1}{2}\angle COA + \angle ACO = 90^\circ\end{aligned}$$

and $DE \perp OC$ as desired.

Problem 15

We have an $n \times n$ table ($n > 100$) in which $n - 1$ entries are 1 and the rest are 0. We may choose an entry, subtract 1 from it, and add 1 to the other entries in its row and column. By this process, can we make all of the entries of the table equal?

Solution. The answer is no. We look at all the numbers modulo 3. Let a_{ij} be the number in row i, column j. We claim that $C = a_{ij} - a_{kj} + a_{km} - a_{im}$ remains constant modulo 3. (If one of the 4 numbers is chosen, then the value of C will increase or decrease by 3; otherwise, it will remain the same.)

Now, since there are $n-1$ ones, one row w has all zeroes and another row x has between 1 and $n-1$ ones. So, there are four entries $a_{wy}=0$, $a_{xy}=0$, $a_{xz}=1$, $a_{wz}=0$. From above, $a_{wy}-a_{xy}+a_{xz}-a_{wz}$ must always be congruent to 1 modulo 3; then these four entries can never be equal or else $a_{wy}-a_{xy}+a_{xz}-a_{wz}$ would be divisible by 3, a contradiction. Therefore, the entries of the table can never become equal, as claimed.

Problem 16

Find all ways to distribute the numbers from 1 to 9 in a 3×3 table so that for each of the 6 squares formed by the entries of the table, the numbers at the corners of the square have the same sum.

Solution. The only distribution (with respect to reflections and rotations) is

$$\begin{bmatrix} a_{11} & a_{12} & a_{13} \\ a_{21} & a_{22} & a_{23} \\ a_{31} & a_{32} & a_{33}, \end{bmatrix} = \begin{bmatrix} 1 & 6 & 7 \\ 8 & 5 & 2 \\ 3 & 4 & 9 \end{bmatrix}.$$

Let k be the common sum. Summing the entries in each of the four corner 2×2 squares, we have

$$\begin{aligned} 4k &= (a_{11}+a_{13}+a_{33}+a_{31})+2(a_{12}+a_{23}+a_{32}+a_{21})+4a_{22} \\ &= 3k+4a_{22} \end{aligned}$$

Summing the entries from the other two of the six squares, we have

$$\begin{aligned} 2k &= (a_{11}+a_{13}+a_{33}+a_{31})+(a_{12}+a_{23}+a_{32}+a_{21}) \\ &= 45-a_{22}, \end{aligned}$$

which leads to $a_{22}=5$, $k=20$.

Now, any square with both 1 and 5 must have two unused numbers adding to 14; the only such numbers are 6 and 8. Then 1 can be in at most one square with 5, so it must be in a corner and next to 6 and 8. Similarly, 3 must be in a corner and next to 4 and 8. Thus, 1, 8, and 3 are on one side of the 9×9 square in that order next to 6, 5, and 4, and it follows that the final three numbers must be 7, 2, and 9, giving the presented square.

Problem 17

A group of shepherds have 128 sheep among them. If one of them has at least half of the sheep, each other shepherd steals as many sheep as he already has. If two shepherds each have 64 sheep, one of these two shepherds steals all the sheep from the other. Suppose seven rounds of theft occur. Prove that one shepherd ends up with all of the sheep.

Solution. We prove by induction that after i rounds of theft, $0 \le i \le 7$, the number of sheep belonging to each shepherd is divisible by 2^i. For the base case, after 0 rounds of theft, the number of sheep belonging to each shepherd is divisible by 1. Now suppose the claim is true for $i = k \le 6$—that is, after k rounds of theft, the number of sheep belonging to each shepherd is divisible by 2^k. After another round of theft, the thieves double their sheep so the number of sheep each thief has is divisible by 2^{k+1}. Since the total number of sheep, 128, is also divisible by 2^{k+1}, the number of remaining sheep—belonging to the victim—is also divisible by 2^{k+1}. This proves the claim for $i = k+1$; by induction, it is true for all i, $0 \le i \le 7$.

Then after 7 rounds of theft, the number of sheep each shepherd has is divisible by 128. Some shepherd must have at least one sheep, so he must have all 128, as desired.

Problem 18

Let O be the circumcenter of acute triangle ABC. Let S_A, S_B, S_C be the circles centered at O tangent to BC, CA, AB, respectively. Show that the sum of the angles between the tangents to S_A at A, the tangents to S_B at B, and the tangents to S_C at C is $180°$.

Solution. Let $R = OA = OB = OC$ be the circumradius of triangle ABC; let r_A be the radius of S_A; let 2α be the angle between the tangents to S_A from A; let $\angle OCB = \angle OBC = \alpha'$. Then

$$2\alpha = 2\sin^{-1}\frac{r_A}{OA} = 2\sin^{-1}\frac{r_A}{OB} = 2\alpha' = 180° - 2\angle A.$$

Similarly, the other two angles are $180° - 2\angle B$, $180° - 2\angle C$, respectively. The desired result follows easily.

Problem 19

In the City Council elections, each voter, if he chooses to participate, votes for himself (if he is a candidate) and for each of his friends. A pollster predicts the number of votes each candidate will receive. Prove that the voters can conspire so that none of the pollster's numbers are correct.

Solution. We prove the claim by construction, showing we can change our "roster" of who votes step by step until none of the pollster's numbers are correct. Begin with an empty roster. At each step, find any candidate who would receive the predicted number of votes if the current roster were used; then add him to the roster.

Suppose the pollster predicted a candidate would get k votes. If the candidate is added to the roster during a step, he gets at least $k+1$ votes in every roster from that step on; so, no future roster would make him receive k votes and he cannot be added again.

Then in each step, at least 1 more prediction is foiled and will remain incorrect; so if there are n candidates, this process terminates in at most n steps. When the process does terminate, there are no duplicate names on the roster (from above); and all the predictions are wrong by construction, as desired.

Problem 20

The roots of two quadratic polynomials are negative integers, and they have one root in common. Can the values of the polynomials at some positive integer be 19 and 98?

Solution. The answer is no. Since the polynomials have one negative root in common, they share a common factor $x + a$ for some positive integer a. Then their values on a positive integer n must share a common factor $n + a \geq 1 + 1 = 2$. However, 19 and 98 do not share any such common factor.

Problem 21

On a pool table that is in the shape of a regular 1998-gon $A_1A_2 \cdots A_{1998}$, a ball is shot from the midpoint of A_1A_2. It bounces off sides $A_2A_3, \ldots, A_{1998}A_1$ in succession (such that the angle of approach equals the angle of departure) and ends up where it started. Prove that the trajectory must be a regular 1998-gon.

Solution. If the ball is shot at the midpoint of A_2A_3, the ball will hit the midpoints of each side, ending up where it started and forming a regular 1998-gon. We show that this path is unique.

Let N_1 be the original 1998-gon and let N_i be the reflection of N_{i-1} across A_iA_{i+1} (where $A_{1999} = A_1$). Then the path of the ball maps onto a straight line through these 1998-gons. For the ball to end up where it started, this line must pass through the midpoint of A_1A_2 in N_{1998}. The given path does exactly that; any other path maps to a different line and cannot pass through this midpoint. So, the path is unique, as claimed.

Problem 22

Find all real x such that $\{(x+1)^3\} = x^3$, where $\{x\} = x - \lfloor x \rfloor$ is the fractional part of x.

Solution. Note that $-1 < x < 1$. If $x < 0$, then

$$x^3 = \{(1+x)^3\} = (1+x)^3$$

which has no real solution.

If $x \geq 0$, then $x^3 = \{(x+1)\}^3 = \{x^3 + 3x(1+x)\}$. So $3x(1+x)$ is an integer; solving $3x(1+x) = k$ for $k = 0, 1, \ldots, 5$, we obtain $x = (\sqrt{9+12k} - 3)/6$ for $k = 0, 1, \ldots, 5$ as all the solutions of the problem.

Problem 23

In the pentagon $A_1A_2A_3A_4A_5$ are drawn the bisectors ℓ_1, ℓ_2, ℓ_3, ℓ_4, ℓ_5 of the angles A_1, A_2, A_3, A_4, A_5, respectively. Bisectors ℓ_1 and ℓ_2 meet at B_1, ℓ_2 and ℓ_3 meet at B_2, and so on. Can the pentagon $B_1B_2B_3B_4B_5$ be convex?

Solution. The answer is no. Each B_i is the center of a circle tangent to $A_{i-1}A_i$, A_iA_{i+1}, and $A_{i+1}A_{i+2}$; WLOG let B_1 correspond to the smallest of these circles, so that $A_1B_1 < A_1B_5$ and $A_2B_1 < A_2B_2$. If the pentagon is convex, all B's lie inside angle $B_5B_1B_2$, which is seen to force $A_3B_2 > A_3B_3$ and $A_5B_5 > A_5B_4$. Again by convexity, all B's must lie inside angle $B_1B_2B_3$ which forces $A_4B_3 < A_4B_4$. And all B's must lie inside angle $B_4B_5B_1$, giving $A_4B_4 < A_4B_3$. Contradiction.

Problem 24

A cube of side length n is divided into unit cubes by partitions (each partition separates a pair of adjacent unit cubes). What is the smallest number of partitions that can be removed so that from each cube, one can reach the surface of the cube without passing through a partition?

Solution. The answer is $(n-2)^3$. Consider the inner $(n-2)^3$ unit cubes as the vertices of a graph, where adjacent vertices represent two unit cubes where a partition has been removed. Also include a "free" vertex representing surface unit cubes. We must connect all $(n-2)^3+1$ vertices; to do this, we need at least $(n-2)^3$ edges. In other words, we must remove at least $(n-2)^3$ partitions; and by removing the top partition from each of the inner $(n-2)^3$ unit cubes, we achieve this minimum.

Problem 25

I choose a number from 1 to 144, inclusive. You may pick a subset of $\{1, 2, \ldots, 144\}$ and ask me whether my number is in the subset. An answer of "yes" will cost you 2 dollars, an answer of "no" only 1 dollar. What is the smallest amount of money you will need to be sure to find my number?

Solution. The answer is 11 dollars. Let F_n be the nth Fibonacci number and let $f(n)$ be the minimum amount of money needed to guarantee finding a number using the above rules from a set of $n \geq 1$ numbers. Note that f is clearly nondecreasing; also, if we first ask about a subset of m numbers, we could spend up to

$$\max(f(m)+2, f(n-m)+1)$$

dollars.

We first prove by induction on n that $f(F_{n+1}) \leq n$. The claim is obvious for $n = 0$ and $n = 1$; now suppose it is true for all $k < n$, $n \geq 2$. Then with a set of F_{n+1} numbers, we first name a subset with F_{n-1} elements. Then we spend up to

$$\max(f(F_{n-1})+2, f(F_{n+1}-F_{n-1})+1) \leq \max(n, n) = n$$

dollars, as claimed.

Next we prove by induction on x that if $F_n < x \leq F_{n+1}$, then $f(x) \geq n$. For $x = 1$, $f(1) \geq 0$; and for $x = 2$, $f(2) \geq 2$. Now assume the claim is true for all $k < x$, where $x \geq 3$ and $F_n < x \leq F_{n+1}$.

If we first give a subset with at most F_{n-2} elements, then we could pay at least

$$f(x-F_{n-2})+1 \geq f(F_{n-1}+1)+1 \geq n$$

dollars. If we first give a subset with at least $F_{n-2}+1$ elements, then again we could pay at least

$$f(F_{n-2}+1)+2 \geq n$$

dollars. Either way, we could be forced to spend at least n dollars, as claimed.

Setting $n = 11$ in both results, we can definitely find the number with 11 dollars, but not necessarily with less.

Problem 26

A positive integer is written on a board. We repeatedly erase its unit digit and add 5 times that digit to what remains. Starting with 7^{1998}, can we ever end up at 1998^7?

Solution. The answer is no. Let a_n be the n-th number written on the board; let u_n be the unit digit and $a_n = 10t_n + u_n$. We have

$$a_{n+1} = t_n + 5u_n \equiv 50t_n + 5u_n \equiv 5(10t_n + u_n) = 5a_n \pmod{7}.$$

Since $a_1 = 7^{1998} \equiv 0 \not\equiv 1998^7 \pmod{7}$, we can never obtain 1998^7 from 7^{1998}.

Problem 27

On an infinite chessboard, we draw a polygon with sides along the grid lines. A unit segment along the perimeter is colored black or white according to whether it touches a black or white square inside the polygon. Let A, B, a, b be the numbers of black segments, white segments, black squares inside the polygon, and white squares inside the polygon, respectively. Show that $A - B = 4(a - b)$.

Solution. Number each square in the polygon $1, 2, \ldots$, and let c_i be the number of squares in the polygon bordering square i. The there are $4 - c_i$ colored segments bordering square i, and they are the same color as the square. Note that $\sum_{black} c_i = \sum_{white} c_i$ since every time a black square borders a white square, the white square borders the black square. Then

$$A - B = \sum_{black} (4 - c_i) - \sum_{white} (4 - c_i) = 4(a - b),$$

as desired.

Problem 28

A sequence $\{a_n\}_{n=1}^{\infty}$ of positive integers contains each positive integer exactly once. Moreover, for every pair of distinct positive integers m and

n,

$$\frac{1}{1998} < \frac{|a_n - a_m|}{|n - m|} < 1998.$$

Show that $|a_n - n| < 2000000$ for all n.

Solution. Consider the positive integers as blocks in a sidewalk from left to right, and consider $a_1, a_2, \ldots$ as the sequence of blocks we visit; we must visit every block exactly once.

First let $n = m + 1$ in the given inequality. From the RHS inequality, we have $|a_{m+1} - a_m| < 1998$, i.e., we can move at most 1997 blocks on each move.

Then let $a_n = a_m + 1$ in the given inequality. From the LHS inequality, we have $|n - m| < 1998$. i.e., we can take at most 1997 moves from one block to an adjacent block.

We claim that if we are on block a_i, every unvisited number to our left is greater than $a_i - 2000000$. Assume the opposite for the sake of contradiction, and say we have not visited $m \leq a_i - 2000000$. Let a_j be the rightmost block with $j \geq i$ that we visit before we visit block m. From a_j to m, we make at least $2000000/1997 > 1000$ moves. By the definition of j, after reaching m we still must visit $a_j + 1$. This takes at least another 1000 moves. So, it takes at least 2000 moves to go from a_j to $a_j + 1$, a contradiction. Thus our assumption is wrong and our claim is correct.

Now, suppose that $a_n - n \geq 2000000$. When we visit a_n, we've only visited n blocks. So, there is a block between 1 and n that we still must visit. But this is at least 2000000 blocks to the left of a_n, a contradiction.

On the other hand, suppose that $n - a_n \geq 2000000$. By the time we visit a_n, there must be a block $a_m > n$ we already visited. But then when we visited a_m, we hadn't visited a_n more than 2000000 blocks to our left, again a contradiction.

Thus, $|a_n - n| < 2000000$, as desired.

Problem 29

There are two arcs of the parabola $y = x^2 + px + q$, each sandwiched between the rays $y = x$ and $y = 2x$ ($x \geq 0$). These two arcs are projected onto the x axis. Prove that the right projection has length 1 more than the left projection.

Solution. Let a, b with $a \leq b$ be the roots of $x = x^2 + px + q$. Then $a + b = 1 - p$. Let c, d with $c < d$ be the roots of $2x = x^2 + px + q$. Then $c + d = 2 - p$. So $(d - b) = (a - c) + 1$, as desired.

Problem 30

A convex polygon is partitioned into parallelograms. Prove that there are at least three vertices contained in only one parallelogram.

Solution. It is simple to verify that the polygon must have an even number of sides, that the polygon must be centrally symmetric, and that any side of a parallelogram must be parallel to to a side of the polygon.

Let the polygon have $2n$ sides. We prove that among any $n-1$ consecutive vertices, one of them is contained in only one parallelogram; from this fact, the desired result follows. Suppose the opposite for the sake of contradiction; label the vertices of the polygon $P_1, P_2, \ldots, P_{2n}$ counterclockwise, and suppose without loss of generality that none of $P_1, P_2, \ldots, P_{n-1}$ is contained in exactly one parallelogram. Let $\theta_1, \theta_2, \ldots, \theta_{2n}$ be the counterclockwise angles between vector $\overrightarrow{P_{2n}P_1}$ and vectors $\overrightarrow{P_1P_2}$, $\overrightarrow{P_2P_3}$, …, $\overrightarrow{P_{2n}P_1}$ (where $0 < \theta_i \leq 2\pi$).

Now, each of the angles at $P_1, P_2, \ldots, P_{n-1}$ is divided by a side P_iQ_i of some parallelogram. Let ϕ_i equal the counter-clockwise angle between vectors $\overrightarrow{P_{2n}P_1}$ and $\overrightarrow{P_iQ_i}$ (again, with $0 < \phi_i \leq 2\pi$). We prove that $\phi_1, \phi_2, \ldots, \phi_{n-1}$ are non-increasing. Suppose otherwise, i.e., that $\phi_j < \phi_{j+1}$ for some $j, 1 \leq j \leq n-2$. Look at the parallelograms along side P_jP_{j+1}; looking at them as they come from vertex P_j, the angles their sides (not parallel to $\overrightarrow{P_jP_{j+1}}$) make with $\overrightarrow{P_{2n}P_1}$ are non-increasing, so they are less than or equal to ϕ_j. On the other hand, looking at the parallelograms as they come from vertex P_{j+1}, the angles their sides (not parallel to $\overrightarrow{P_jP_{j+1}}$) make with $\overrightarrow{P_{2n}P_1}$ are non-decreasing, so they are at least $\phi_{j+1} > \phi_j$, a contradiction. Thus, the ϕ_i are non-increasing.

But the sides of the parallelograms are parallel to the sides of the polygon, so each ϕ_i must equal to some θ_j; in particular, $\phi_1 \leq \theta_{n-1}$ and $\phi_{n-1} \leq \phi_1 \leq \theta_{n-1}$, which is impossible. Thus our original assumption was false.

Problem 31

Let $S(x)$ be the sum of the digits in the decimal representation of the number x. Do there exist three natural numbers a, b, c such that $S(a+b) < 5$, $S(b+c) < 5$, $S(c+a) < 5$, but $S(a+b+c) > 50$?

Solution. The answer is yes. It is easier to find $a+b$, $b+c$, $c+a$ instead. Since $a+b+c$ is an integer, their sum $2(a+b+c)$ must be even; since a, b, c are positive, they must satisfy the triangle inequality. Finally, $a+b+c$ must have a digit sum of at least 51.

This leads to the solution $a+b = 100001110000$, $b+c = 11110000000$, $c+a = 100000001110$. These four numbers have digit sum 4, and $a+b+c = 105555555555$ has digit sum 51. And we verify that a, b, c are indeed integers:

$$a = 105555555555 - 11110000000 = 94445555555$$

$$b = 105555555555 - 100000001110 = 5555554445$$

$$c = 105555555555 - 100001110000 = 5554445555.$$

Problem 32

A maze is an 8×8 board with some adjacent squares separated by walls, such that there is a path from any square to any other square not passing through a wall. Given the command LEFT, RIGHT, UP or DOWN, a pawn advances one square in the corresponding direction if this movement is not blocked by a wall or an edge of the board, and otherwise does nothing. God writes a program (a finite sequence of such commands) and gives it to the Devil, who then constructs a maze and places the pawn on one of the squares. Can God ensure that the pawn will land on every square of the board no matter what the Devil does?

Solution. The answer is yes. The Devil creates a maze and chooses the pawn's starting position; together, call these the Devil's *arrangement*. There are only finitely many arrangements the Devil can create: say, N. We can clearly create a program that works in one of them. Then while our program still does not work for all N arrangements, choose an arrangement X it does not work for.

In this arrangement X, the program moves the pawn to some location; simply add more commands so that the pawn then travels to all the squares in X. In this manner, we can create a program that works for all N arrangements, as claimed.

Problem 33

I set 5 watches, then you move some of them forward so that they show the same time, and add up the intervals through which you moved each watch.

How large can I force this sum to be, no matter how you manipulate the watches?

Solution. We assume the watches cover the 24 hours in a day. Then you can force the sum to be 48 hours, but no greater. We claim that we can always use at most 48 hours by setting the watches forward to one of the original watch times. Order the original times and let a, b, c, d, e be the periods between consecutive times. Then moving the watches forward to an original watch time will move them forward through a total of $a + 2b + 3c + 4d$, $2a + 3b + 4c + e$, $\ldots$, or $b + 2c + 3d + 4e$ hours. These five possibilities sum to $10(a + b + c + d + e) = 240$, so one of them is at most $240/5 = 48$. And it is easy to see that the maximum is attainable when $a = b = c = d = e = 24/5$.

Problem 34

In triangle ABC, with $AB > BC$, BM is a median and BL an angle bisector. The line through M parallel to AB meets BL at D, and the line through L parallel to BC meets BM at E. Prove that $ED \perp BL$.

First Solution. Let C' be the reflection of C with respect to BL. Let CC' meet BL and BM at D' and E' respectively. Since $E'D' \perp BL$, the desired result follows easily from the claim that $D = D'$ and $E = E'$.

Since $BC = BC'$, triangle BCC' is isosceles so that BD' bisects $\angle CBC'$. Because BD' also bisects $\angle CBA$, C' must be on line AB. Therefore MD' is the midline of triangle ACC', $MD' \parallel AB$, and $D = D'$.

Let $AB = c, BC = a, CA = b$. From the angle bisector theorem,

$$LC = \frac{ab}{a+c}, \qquad \frac{ML}{LC} = \frac{\frac{b}{2} - \frac{ab}{a+c}}{\frac{ab}{a+c}} = \frac{c-a}{2a}.$$

Since $MD' \parallel AB$, triangles $BC'E'$ and $MD'E'$ are similar. We have

$$\frac{ME'}{E'B} = \frac{MD'}{C'B} = \frac{\frac{c-a}{2}}{a} = \frac{c-a}{2a} = \frac{ML}{LC}.$$

Therefore $LE' \parallel BC$ and $E' = E$, as claimed.

Second Solution. Let EL and DM meet at X. We have $\angle XDL = \angle ABL = \angle LBC = \angle DLX$, so $DX = LX$.

Also, since M is the midpoint of AC, MD must intersect BC at its midpoint N. But then triangles MEL and MBC are homothetic, so X is homothetic to N and must be the midpoint of EL. Thus, $EX = LX = DX$.

Then EL is a diameter of the circle centered at X with radius XD, which implies that $\angle EDL = 90^\circ$, as desired.

Problem 35

A jeweler makes a chain from $N > 3$ numbered links for a mischievous customer, who then asks the jeweler to change the order of the links, in such a way as to maximize the number of links the jeweler must open. How many links had to be opened?

Solution. The answer is $\lfloor \frac{3N}{4} \rfloor$. To rearrange the links, it suffices to open enough links so that no two closed links are adjacent either in the original arrangement, or in the desired arrangement. To do so, open every other link in the original arrangement for a total of at most $\lfloor \frac{N}{2} \rfloor$ links. Now, there still might be sequences of adjacent, still-closed links in the desired arrangement. In each of these sequences, open every other link. Then we can rearrange the links by opening at most $\lfloor \frac{N}{2} \rfloor + \lfloor \lceil \frac{N}{2} \rceil / 2 \rfloor = \lfloor \frac{3N}{4} \rfloor$ links.

On the other hand, the jeweler can be forced to open $\lfloor \frac{3N}{4} \rfloor$ links. Let $l = \lfloor \frac{N}{4} \rfloor$. If we number the links in the original chain $1, 2, \ldots, N$, then suppose we wish to obtain the chain

$$4l + 2, 3, 1, 4, 2, 7, 5, 8, 6, \ldots, 4l - 1, 4l - 3, 4l, 4l - 2, 4l + 1, 4l + 3$$

(of course, the links numbered $4l + 1, 4l + 2, 4l + 3$ might not exist). Then in each of the sets $\{1, 2, 3, 4\}, \{5, 6, 7, 8\}, \ldots, \{4l - 3, 4l - 2, 4l - 1, 4l\}, \{4l + 1, 4l + 2, 4l + 3\}$, at most one link can remain unopened. So, at least $N - \lceil \frac{N}{4} \rceil = \lfloor \frac{3N}{4} \rfloor$ links must be opened.

Problem 36

Two positive integers are written on the board. The following operation is repeated: if $a < b$ are the numbers on the board, then a is erased and $ab/(b - a)$ is written in its place. At some point the numbers on the board are equal. Prove that again they are positive integers.

First Solution. Call the original numbers x and y and let $L = \text{lcm}(x, y)$. For each number n on the board consider the quotient L/n; during each operation, the quotients L/b and L/a become L/b and $L/a - L/b$. This is the Euclidean algorithm, so the two equal quotients would be $\gcd(L/b, L/a)$ and the two equal numbers on the board are $L/\gcd(L/x, L/y)$. But

$\gcd(L/x, L/y) = 1$, because otherwise x and y would both divide $L/\gcd(L/x, L/y)$ and L would not be a least common multiple. So, the two equal numbers equal $L = \operatorname{lcm}(x, y)$, an integer.

Second Solution. Again, let x and y be the original numbers and suppose both numbers eventually equal N. We prove by induction, on the number of steps k before we obtain (N, N), that all previous numbers divide N. Specifically, $x \mid N$, so N must be an integer.

The claim is clear for $k = 0$. Now assume that k steps before we obtain (N, N), the numbers on the board are $(c, d) = (N/p, N/q)$ for some integers $p < q$. Then reversing the operation, the number erased in the $(k+1)$st step must be $cd/(c+d) = N/(p+q)$ or $cd/(c-d) = N/(q-p)$, completing the inductive step.

Problem 37

Two lines parallel to the x-axis meet the graph

$$y = ax^3 + bx^2 + cx + d$$

in the points A, D, E and B, C, F, respectively, in that order from left to right. Prove that the length of the projection of the segment CD onto the x-axis equals the sum of the lengths of the projections of AB and EF.

Solution. WLOG, let $a > 0$, and let the lines be $y = y_1$ and $y = y_2$, with $y_1 < y_2$. Also define a_i so that $A = (a_1, y_1)$, $B = (a_2, y_2)$, $C = (a_3, y_2)$, $D = (a_4, y_1)$, $E = (a_5, y_1)$, $F = (a_6, y_2)$. Then since the function is a cubic, $a_i < a_j \iff i < j$. Since a_1, a_4, a_5 are the roots of $ax^3+bx^2+cx+d-y_1$ and a_2, a_3, a_6 are the roots of $ax^3+bx^2+cx+d-y_2$, we have $a_1 + a_4 + a_5 = -b/a = a_2 + a_3 + a_6$, which implies the result.

Problem 38

Two polygons are given such that the distance between any two vertices of one polygon is at most 1, but the distance between any two vertices of different polygons is more than $1/\sqrt{2}$. Prove that the polygons have no common interior point.

Solution. Assume the contrary for a contradiction. Then some edge A_1A_2 of polygon P_1 intersects an edge B_1B_2 of the other polygon P_2. A_1 and A_2 are not on B_1B_2, because then the distance between two vertices of different polygons would be less than $1/2$. A similar statement holds about

B_1 and B_2, so quadrilateral $A_1B_1A_2B_2$ is convex. Applying the law of cosines to triangle $A_1B_1A_2$ and using the constraints of the problem, we have $\cos\angle A_1B_1A_2 \geq 0 \Rightarrow \angle A_1B_1A_2 \leq 90°$. The same is true about the other three angles, so in order for the angles of the quadrilateral to sum to $360°$, all angles must be $90°$. It follows from the Pythagorean Theorem that $(A_1A_2)^2 = (A_1B_1)^2 + (B_1A_2)^2 > 1$. But we also know that $A_1A_2 \leq 1$, a contradiction.

Problem 39

Let D, E, F be the feet of the angle bisectors of angles A, B, C, respectively, of triangle ABC, and let K_a, K_b, K_c be the points of contact of the tangents to the incircle of ABC through D, E, F (that is, the tangent lines not containing sides of the triangle). Prove that the lines joining K_a, K_b, K_c to the midpoints of BC, CA, AB, respectively, pass through a single point on the incircle of ABC.

Solution. Let A_1, B_1, C_1 be the midpoints of BC, CA, AB, respectively. We claim that triangles $A_1B_1C_1$ and $K_aK_bK_c$ are homothetic. Then note that the circumcircle of $K_aK_bK_c$ is incircle ω_1 of triangle ABC, and that the circumcircle of $A_1B_1C_1$ is the nine-point circle ω_2 of triangle ABC. From Feuerbach's theorem, ω_1 and ω_2 are tangent; let K be the point of tangency. Then K is the center of the homothety and A_1K_a, B_1K_b, C_1K_c are concurrent at K.

To prove that these two triangles are homothetic, it suffices to prove that $K_aK_b \parallel A_1B_1$, $K_bK_c \parallel B_1C_1$, $K_cK_a \parallel C_1A_1$. We prove the first relation; the others follow similarly.

Let ω_1 touch BC, CA, AB at A_2, B_2, C_2, respectively. Extend K_aK_b to meet AC and BC at B' and A' respectively. WLOG, let $\angle A \geq \angle B \geq \angle C$. Then points B_2, K_b, K_a, A_2 are on ω in that order, and:

$$\begin{aligned} \angle B_2EK_b &= 2\angle B_2EB &= \angle B + 2\angle C, \\ \angle K_aDA_2 &= 2\angle ADA_2 &= \angle A + 2\angle C, \end{aligned}$$

and minor arcs

$$\begin{aligned} \overset{\frown}{B_2A_2} &= \angle A + \angle B, \quad \overset{\frown}{B_2K_b} &= \angle A - \angle C, \\ \overset{\frown}{K_aA_2} &= \angle B - \angle C, \quad \overset{\frown}{K_bK_a} &= 2\angle C. \end{aligned}$$

Also notice that B, A_2, D, A_1, C are on BC in that order. Since DK_a is tangent to ω_1 and K_b is not on $\overset{\frown}{K_aA_2}$, A' and A_2 are on different sides of AD. Therefore $\angle CA'K_a$ is the exterior angle of triangle $A'K_aD$ and

$$\angle CA'K_a = \angle DK_aA' + \angle CDK_a = \frac{\overset{\frown}{K_bK_a}}{2} + \overset{\frown}{K_aA_2} = \angle B.$$

So, $K_aK_b \parallel A_1B_1$, as claimed.

Problem 40

Certain subsets of a given set are distinguished. Each distinguished subset contains $2k$ elements, where k is a fixed positive integer. It is known that given a subset with at most $(k+1)^2$ elements, either it contains no distinguished subset, or all of its distinguished subsets have a common element. Show that all of the distinguished subsets have a common element.

Solution. We induct on the number of distinguished subsets. If there is only one, the problem is trivial; if there are two, then their union has at most $4k \le (k+1)^2$ elements, so they share a common element.

Now, suppose every n distinguished subsets have a common element. Then given $n+1$ distinguished subsets $D_1, \ldots, D_n$, every n of them have a common element c_i. If any of these common elements equal each other, we are done; otherwise, there are at least $n+1$ distinct c_i, each in exactly n of the D_i.

Since there are $2k$ elements in each D_i, there are at most $2k(n+1)$ elements in $U = D_1 \cup D_2 \cup \ldots \cup D_{n+1}$. But each of the c_i is over-counted $n-1$ times, so there are at most

$$2k(n+1) - (n+1)(n-1) = 2kn - n^2 + 2k + 1 \le (k+1)^2$$

elements in U, which implies the inductive step.

Problem 41

The numbers 19 and 98 are written on a board. Each minute, each number is either incremented by 1 or squared. Is it possible for the numbers to become identical at some time?

Solution. The answer is no. We prove so by contradiction. Let (a_i, b_i) denote the numbers on the board after ith minute, so that $(a_0, b_0) = (19, 98)$ or $(98, 19)$. Suppose, on the contrary, that somehow $(a_{n+1}, b_{n+1}) = (x, x)$ can be obtained. We further suppose that (x, x) is the pair one can obtain in the minimum amount of steps. WLOG,

$(a_n, b_n) = (x-1, \sqrt{x})$. It is clear that a_i, b_i are all positive integers for all i (in fact, a_i, $b_i \geq 19$). So $x = c^2$ is a perfect square and $(a_n, b_n) = (c^2-1, c), c \geq 19$. Since b_0 is positive, $n < c < 2c-2$; but then since the largest perfect square less than a_n is $a_n - (2c-2)$, the last n steps on a_i must $+1$.

Thus, a_0 is bigger than b_0 and a_{n+1} is the smallest square larger than a_0. So, $a_0 = 98$, $a_{n+1} = 100$, and $n+1 = 2$. But on the other hand, it is impossible to obtain 100 from 19 in only 2 steps, a contradiction. So it is impossible for the numbers to become identical at some time.

Problem 42

A binary operation $*$ on real numbers has the property that $(a * b) * c = a + b + c$. Prove that $a * b = a + b$.

Solution. First we note that if $a * b = a * d$, then

$$a + b + c = (a * b) * c = (a * d) * c = a + d + c$$

so that $b = d$. Similarly, if $a * b = d * b$, then $a = d$. We also note that $a * b = b * a$; to show this, let $a * b = d_1$, $b * a = d_2$. Then $d_1 * c = a + b + c = d_2 * c$ and $d_1 = d_2$.

Now, let $a * 0 = x$, so that $x * 0 = (a * 0) * 0 = a$. So

$$2x = (x * 0) * x = a * x = x * a = (a * 0) * a = 2a,$$

implying that $x = a$ and $a * 0 = a$. Now $a + b = (a * b) * 0 = a * b$, as desired.

Problem 43

A convex polygon with $n > 3$ vertices is given, no four vertices lying on a circle. We call a circle circumscribed if it passes through three vertices and contains all of the others. A circumscribed circle is called a boundary circle if it passes through three consecutive vertices of the polygon, and it is called an inner circle if it passes through three pairwise non-consecutive vertices. Prove that the number of boundary circles is 2 more than the number of inner circles.

First Solution. Let us number the vertices of the polygon counterclockwise $a_1, a_2, \ldots, a_n$. A segment of vertices is a sequence of consecutive vertices $[a_k, a_{k+s}] = \{a_k, a_{k+1}, \ldots, a_{k+s}\}$. (Subscripts are modulo n.) A segment of vertices is called empty if $s = 1$. For example, $[a_n, a_1]$ is an empty segment (but $[a_1, a_n]$ is not). We will call a triangle circumscribed, boundary, or inner if its circumcircle is circumscribed, boundary, or inner, respectively. We will write the vertices of a triangle in the counterclockwise direction, so triangle abc means c is in $[b, a]$ and not in $[a, b]$. We will call $[a, b]$, $[b, c]$, $[c, a]$ the segments of triangle abc.

Lemma 22. *If abc and klm are circumscribed triangles, then their interiors do not intersect. More precisely, the triangles can intersect only in an edge or a vertex.*

Proof. Suppose ab intersects kl in a point x. Consider the convex quadrilateral $akbl$. There exists a circle passing through a and b such that k and l are inside (namely, the circumcircle of abc), so $\angle akb + \angle alb > \pi > \angle kal + \angle kbl$. Similarly, there exists a circle through k and l containing a and b, so $\angle kal + \angle kbl > \pi > \angle akb + \angle alb$, a contradiction. We cannot have one of the triangles abc, klm inside the other because a, b, c, k, l, m are the vertices of a convex polygon. Hence the interiors of abc and klm do not intersect. □

Lemma 23. *Suppose either $[a, b]$ is empty or there exists d such that adb is circumscribed. Assume $[b, a]$ is not empty. Then there exists exactly one vertex $c \in [b, a]$ such that abc is circumscribed.*

Proof. In fact, c is the vertex in $[b, a]$ minimizing $\angle bca$ (there is only one such vertex because no four vertices are concyclic). Then all other vertices in $[b, a]$ are inside circle abc. If c' is any other vertex in $[b, a]$, then c lies outside circle abc', so c is uniquely determined. It remains to show that every vertex in $[a, b]$ besides a and b lies inside circle abc. If $[a, b]$ is empty, there is nothing to prove, so assume there is a circumscribed triangle adb. Since c lies inside circle adb, $\angle acb + \angle adb > \pi$. For $x \in [a, b]$ distinct from a and b, x lies inside circle adb so $\angle axb > \angle adb$. Hence $\angle acb + \angle axb > \angle acb + \angle adb > \pi$ and x lies inside circle abc. So abc is circumscribed. □

Now we begin the proof of the main result. Let N be the number of inner triangles; draw all of their edges. We first consider the case $N > 0$. By the first lemma the interiors of these triangles do not intersect, so the vertices of these triangles divide $a_1, a_2, \ldots, a_n$ into segments of two types:

(1) segments $[a, b]$ of an inner triangle, which contain no other segments of inner triangles;

(2) segments $[d, e]$ which contain no segments of inner triangles, but $[f, d]$ and $[e, g]$ are segments of inner triangles for some f and g.

We call these segments i-segments. There are $N+2$ i-segments of the first type. This is obvious when $N = 1$, and if we add inner triangles one at a time, each new triangle replaces a segment of type 1 with two segments of type 1 and possibly some segments of type 2. If abc is any boundary triangle, then a, b, c are in the same i-segment by the first lemma. The statement of the problem follows from the next lemma.

Lemma 24.

(1) Each i-segment of type 1 contains exactly one boundary triangle.

(2) i-segments of type 2 contain no boundary triangles.

Proof. **(1)** We prove a slightly stronger statement: If $[a, b]$ is a nonempty segment of a circumscribed triangle bda and $[a, b]$ contains no inner triangles, then there exists a unique boundary triangle in $[a, b]$. We induct on the number of vertices in $[a, b]$. If $[a, b]$ contains just three vertices a, c, b, then acb is circumscribed by the second lemma and the claim holds. If $[a, b]$ contains $k > 3$ vertices, assume the claim holds for $k-1$. By the second lemma there exists a circumscribed triangle acb. It is not inner by assumption, so either $[a, c]$ or $[c, b]$ is empty; assume WLOG $[c, b]$ is empty. Then $[a, c]$ contains $k - 1 \geq 3$ vertices; by the inductive hypothesis, $[a, c]$ contains a unique boundary triangle xyz. Now if klm is any boundary triangle with vertices in $[a, b]$, then by the first lemma k, l, m are in segment $[a, c]$. Hence xyz is the unique boundary triangle in segment $[a, b]$.

(2) Suppose abc is a boundary triangle in a segment $[d, e]$ of type 2. (Here a and d, c and e are not necessarily distinct.) By the second lemma there exists a circumscribed triangle acx_1.

If acx_1 is inner circle, then $[a, c]$ is an inner segment contained in $[d, e]$, contradiction.

If acx_1 is not an inner circle, then one of $[c, x_1]$ and $[x_1, a]$ is empty. WLOG, suppose $[c, x_1]$ is empty. By the first lemma, x_1 is in $[c, d]$. We apply the second lemma to obtain circumscribed triangle ax_1x_2. If ax_1x_2 is inner, we have a contradiction. If not, we will apply the second lemma again and so on to get bigger and bigger triangles; eventually we will get a vertex x outside $[d, e]$, which (by the first lemma) must be in $[g, f]$. At this point we have a triangle xyz with y and z inside $[d, e]$ but x outside. But x cannot be adjacent to y or z because $[f, d]$ and $[e, g]$ are nonempty (since they are segments of an inner triangle). So xyz is an inner triangle and $[y, z]$ is an inner segment contained in $[d, e]$, a contradiction. □

It only remains to consider the case $N = 0$. Pick an empty segment $[a, b]$; by the second lemma there is a circumscribed circle abc. First, suppose abc is a boundary triangle, say $[b, c]$ is empty. Since $n > 3$, $[c, a]$ is not empty, so by the previous lemma there is a unique boundary triangle in $[c, a]$, giving a total of 2 boundary triangles. If abc is not a boundary triangle, then $[b, c]$ and $[c, a]$ are both nonempty. By the first lemma every boundary triangle lies in either $[b, c]$ or $[c, a]$; by the previous lemma each of these segments contains a unique boundary triangle, again giving 2 boundary triangles, as desired.

Second Solution. We present another method based on the first two lemmas. We claim that the circumscribed triangles trianglulate the polygon; specifically, they cover the polygon and intersect only in edges or vertices. The latter part is given by the first lemma; we will use the second lemma to prove the former part.

Start by applying the lemma to an empty segment $[a, b]$; this gives us a circumscribed triangle abc. I claim that if xyz is a circumscribed triangle, then we can cover the convex hull of segment $[x, z]$ with circumscribed triangles. If $[x, z]$ is empty then xyz is sufficient. Otherwise by the second lemma there is a vertex w in $[z, x]$ such that xzw is circumscribed. Since $[w, z]$ and $[z, x]$ are smaller segments we can cover them with circumscribed triangles in the same way; these triangles together with xzw cover xyz. Applying this to segments $[a, c]$ and $[c, b]$ gives a collection of circumscribed triangles which together with abc cover the entire polygon. Hence we have shown that the circumscribed triangles triangulate the polygon.

Let p, q, r be the number of circumscribed triangles containing 0, 1, 2 edges of the polygon respectively; no triangle can contain 3 edges of the polygon since $n > 3$. Note that p is the number of inner triangles and r is the number of boundary triangles. Since the circumscribed triangles triangulate the polygon, the sum of the angles of these triangles equals the sum of the interior angles of the polygon, so $p + q + r = n - 2$. Also, since each edge belongs to exactly one circumscribed triangle, $q + 2r = n$. Subtracting these equations gives $r = p + 2$, as desired.

Problem 44

Each square of a $(2^n - 1) \times (2^n - 1)$ board contains either $+1$ or -1. Such an arrangement is called successful if each number is the product of its neighbors (squares sharing a common side with the given square). Find the number of successful arrangements.

Solution. There are clearly two successful arrangements when $n = 1$, but only one arrangement for larger n. Suppose for contradiction that for some $n > 1$, there is a successful arrangement that contains a -1; consider such an arrangement for the smallest such n.

If the arrangement is not horizontally symmetrical, then multiply each element by the corresponding element opposite it horizontally; doing so keeps the arrangement successful. At the same time, it makes the arrangement horizontally symmetrical, and some element still equals -1 (if all the elements become 1, then the arrangement must have been horizontally symmetrical to start with).

Similarly, we can make the arrangement vertically symmetrical as well. Now, since the board is horizontally symmetrical, each element in the middle column equals the product of its neighboring elements in the middle column (its horizontal neighbors are equal and contribute 1 to the product). Then the middle row must equal either all $+1$'s or else $-1, -1, 1, -1, -1, 1, \ldots, -1, -1$. The second case is impossible since $2^n - 1$ cannot equal $2 \pmod 3$; so, the middle row must be completely filled with 1's.

Similarly, the middle column must be completely filled with 1's. Then these 1's border four smaller successful arrangements with side lengths $2^{n-1} - 1$. By the definition of n, either they are also completely filled with 1's, or else $n = 2$. But even if $n = 2$, it is simple to check that the smaller arrangements must be filled with 1's. Thus, every entry is $+1$, a contradiction; our original assumption was false, and for $n > 1$, there is exactly one successful arrangement (namely, the one with all $+1$'s).

Problem 45

A family S of equilateral triangles in the plane is given, all translates of each other, and any two having nonempty intersection. Prove that there exist three points such that every member of S contains one of the points.

Solution. Take the lengths of all sides to be 1. Orient the plane so that all triangles have a horizontal side and the third vertex is above the horizontal side; we will call this third vertex the upper vertex of the triangle. Then the sides of the triangles are parallel to three directions: the horizontal direction and two others, α and β, which form 60° angles with the horizontal direction. Note that a point X belongs to triangles whose upper vertices are in the "flipped" triangle with lowest vertex X; thus it suffices to show that we can cover all upper vertices with at most three flipped triangles.

There exist two horizontal lines l, s with distance $\sqrt{3}/2$ such that the upper vertex of every triangle lies in the strip between them; otherwise some two triangles would have empty intersection. Similarly, there exist two lines l_α and s_α parallel to the direction α with distance $\sqrt{3}/2$ such that every upper vertex lies in the strip between l_α and s_α, and two lines l_β and s_β with distance $\sqrt{3}/2$ such that every upper vertex lies in the strip between l_β and s_β.

Thus all upper vertices lie in the intersection of these three strips, which is a polygon P with sides parallel to the sides of the original triangles. Lines l_α, s_α, l_β, s_β form a rhombus $ABCD$ with lowest vertex A, cut by the lower horizontal line at EF and by the upper horizontal line at GH. If EF is on or below A, then GH is on or below BD; thus $P = AGH$ and the flipped triangle with low vertex at A covers P. If EF is above A, then let X, Y, Z be the midpoints of the sides of triangle AEF; it is easy to check that the flipped triangles with low vertices at X, Y and Z cover P. Thus all upper vertices can be covered with at most three flipped triangles, as desired.

Problem 46

There are 1998 cities in Russia, each being connected (in both directions) by flights to three other cities. Any city can be reached by any other city by a sequence of flights. The KGB plans to close off 200 cities, no two joined by a single flight. Show that this can be done so that any open city can be reached from any other open city by a sequence of flights only passing through open cities.

Solution. We begin with some terminology. Define a *trigraph* to be a connected undirected graph in which every vertex has degree at most 3. A *trivalent* vertex of such a graph is a vertex of degree 3. In this wording, the problem becomes: We have a trigraph G with 1998 vertices, all of which are trivalent. We want to remove 200 vertices, no two of which are adjacent, such that the remaining vertices stay connected.

We remove the vertices one at a time. Suppose we have deleted k of the 1998 vertices, no two of which are adjacent, such the trigraph G' induced by the remaining vertices is connected. We will show that if $k < 200$, we can always delete a trivalent vertex of G' such that the graph remains connected. This vertex cannot be adjacent in G to any of the other k deleted vertices, because then its degree in G' would be less than 3. Hence repeating this 200 times gives us the desired set of vertices.

Lemma 25. *Let G be a trigraph such that the removal of any trivalent vertex disconnects G. Then G is planar. Moreover, G can be drawn in such a way that every vertex lies on the "outside" face; in other words, for any point P outside some bounded set, each vertex v of G can be joined to P by a curve which does not intersect any edges of G (except at v).*

Proof. We induct on the number of trivalent vertices of G. If G does not contain any trivalent vertices, then G must be a path or a cycle and the claim is obvious. So suppose G contains $n \geq 1$ trivalent vertices and that every trigraph with fewer trivalent vertices can be drawn as described. If G is a tree the claim is obvious, so suppose G contains a cycle; let $v_1, \ldots, v_k$ $(k \geq 3)$ be a minimal cycle. Let $S = \{v_1, \ldots, v_k\}$, and let $T = \{\, i \mid v_i \text{ is trivalent} \,\}$. ($T$ cannot be empty, because then no v_i would be connected to a vertex of degree 3.) For each $i \in T$, let w_i denote the third vertex which is adjacent to v_i (other than v_{i-1} and v_{i+1}), and let S_i be the set of vertices in G which can be reached from w_i without passing through S. (For $i \notin T$, let $S_i = \emptyset$.) We claim that the sets S, S_1, $\ldots$, S_k partition the vertices of G. First, note that if v is a vertex of G not in S, then there is a shortest path joining v to a vertex v_i of S; the penultimate vertex on this path must be w_i, so $v \in S_i$. Now suppose that $v \in S_i \cap S_j$ for some $i \neq j$; then v_i and v_j are trivalent and there exist paths $w_i \rightsquigarrow v$, $w_j \rightsquigarrow v$ which do not pass through S. We will show that there exists a path from every vertex of $G - \{v_i\}$ to v which does not pass through v_i. For $k \neq i$ there is a path $v_k \rightsquigarrow v_j \rightsquigarrow w_j \rightsquigarrow v$; if $w \in S_k$ for $k \neq i$, then there is a path $w \rightsquigarrow w_k \rightsquigarrow v_k \rightsquigarrow v$; if $w \in S_i$, there is a path $w \rightsquigarrow w_i \rightsquigarrow v$. Since $S \cup S_1 \cup \cdots \cup S_k = G$, we have shown that the graph obtained from G by deleting v_i is connected, a contradiction, as v_i is trivalent. Therefore $S_i \cap S_j = \emptyset$ for $i \neq j$. Obviously $S_i \cap S$ is empty for all i; hence S, S_1, $\ldots$, S_k partition the vertices of G. Let G', G_1, $\ldots$, G_k be the induced subgraphs of S, S_1, $\ldots$, S_k in G, respectively. By construction, the only edges in G which are not in one of the graphs G', G_1, $\ldots$, G_k are the edges $v_i w_i$ for $i \in T$. Now G_i is a trigraph with fewer than n trivalent vertices, since at least one of the n trivalent vertices in G is in S. Hence by the inductive hypothesis, we can draw each G_i in the plane in such a way that every vertex lies on the outside face. Since v_1, $\ldots$, v_k was a minimal cycle, there are no "extra" edges between these vertices, so the graph G' is a k-cycle. Now place the vertices of S at the vertices of a small regular k-gon far from all the graphs G_i; then we can draw a curve joining each pair v_i, w_i. It is easy to check that this gives us a drawing of G with the desired properties. □

Now suppose we have removed k vertices from G, no two of which are adjacent, such that the trigraph G' induced by the remaining vertices is connected, and suppose that removing any trivalent vertex of G' disconnects the graph; we must show $k \geq 200$. By the lemma, G' is planar. We will call a face other than the outside one a "proper face." Let F be the number of proper faces of G'; since G' has $1998 - k$ vertices and $2997 - 3k$ edges,

$$F \geq 1 - (1998 - k) + (2997 - 3k) = 1000 - 2k.$$

We now show that no two proper faces can share a vertex. Observe that each vertex belongs to at most as many faces as its degree; thus vertices of degree 1 lie only on the outside face. No two proper faces can intersect in a vertex of degree 2, or that vertex would not lie on the outside face, contradicting the lemma. If two proper faces intersected in a trivalent vertex v, each face would give a path between two of v's neighbors, so removing v would not disconnect the graph, by an argument similar to that of the lemma.

Since each proper face contains at least 3 vertices and no two share a vertex, we have $3F \leq 1998 - k$. Combining this with the previous inequality gives

$$3000 - 6k \leq 3F \leq 1998 - k$$

so $1002 \leq 5k$ and $k \geq 200$, as desired.

Problem 47

The sequence $\omega_1, \omega_2, \ldots$ of circles is inscribed in the parabola $y = x^2$ so that ω_n and ω_{n+1} are externally tangent for $n \geq 1$. Moreover, ω_1 has diameter 1 and touches the parabola at $(0,0)$. Find the diameter of ω_{1998}.

Solution. The answer is 3995.

Let r_n be the radius of ω_n; let $d_n = 2r_n$ be the diameter of ω_n; let $s_n = d_1 + d_2 + \cdots + d_n$; let (x_n, x_n^2) and $(-x_n, x_n^2)$ be the points of tangency of the parabola with ω_n. Finally, we assume that $x_n \geq 0$. Then $(0, s_{n-1} + r_n)$ is the center of ω_n. The slope of the line tangent to the parabola at (x_n, x_n^2) is $2x_n$ and the slope of the normal line is $-1/(2x_n)$.

So

$$\frac{x_n^2 - s_{n-1} - r_n}{x_n} = -\frac{1}{2x_n},$$

$x_n^2 - s_{n-1} - r_n = -1/2$. From distance formula, we have

$$(x_n - 0)^2 + (x_n^2 - s_{n-1} - r_n)^2 = r_n^2,$$

so $x_n^2 = r_n^2 - 1/4$ and $r_n^2 - r_n - s_{n-1} + 1/4 = 0$. Solving the last quadratic equation yields $r_n = (1 + 2\sqrt{s_{n-1}})/2$ and $d_n = 1 + 2\sqrt{s_{n-1}}$. An easy induction shows that $d_n = 2n - 1$ and $d_{1998} = 3995$, as claimed.

Problem 48

The tetrahedron $ABCD$ has all edges of length less than 100, and contains two non-intersecting spheres of diameter 1. Prove that it contains a sphere of diameter 1.01.

Solution. Let $T = ABCD$, S_1, S_2 denote the tetrahedron and the two spheres, respectively. We first apply some transformations to the spheres and the tetrahedron. Since the spheres are non-intersecting, we can move each sphere to a different corner of T so that each sphere is tangent to three faces of T and the spheres are still non-intersecting. Next, we scale down T to $T_1 = A_1B_1C_1D_1$ (with A_1 the image of A and so on), leaving both spheres at their respective corners and with diameter 1, until the spheres are tangent. We claim that there is a sphere S with diameter at least 1.01 inside T' and thus inside T. WLOG, let S_1 and S_2 be in corners A_1 and B_1, respectively. Apply a dilation centered at A_1 with magnitude m such that T_2, the image of T_1, circumscribes S_1. Let B_2 be the image of B_1 under the dilation. Then $A_1B_2/A_1B_1 = m < 1$. Apply translation $\overrightarrow{B_2B_1}$ to T_2 to obtain T_3. Then T_3 circumscribes S_2. So the length of B_2B_1 is equal to the distance between S_1 and S_2, which is 1. Thus $1/m = AB_1/AB_2 > 100/99 > 1.01$. Therefore we can dilate T_2 and S_1 back to T_1 with magnitude $1/m$, and S, the image of S_1, with diameter greater than 1.01, is the inscribed sphere of T_1, as claimed.

Problem 49

A figure composed of 1×1 squares has the property that if the squares of a (fixed) $m \times n$ rectangle are filled with numbers the sum of all of which is positive, the figure can be placed on the rectangle (possibly after being rotated) so that the numbers it covers also have positive sum. (The figure may not be placed so that any of its squares fails to lie over the rectangle.)

Prove that a number of such figures can be put on the $m \times n$ rectangle so that each square is covered by the same number of figures.

Solution. We prove the following generalization:

Let w, w_1, …, w_k be vectors in $\mathbb{R}^N$ with rational entries. Suppose that for every $v \in \mathbb{R}^N$ such that $w \cdot v > 0$, we have $w_i \cdot v > 0$ for some i. Then $w = c_1 w_1 + \cdots + c_k w_k$ for some nonnegative rational numbers c_1, …, c_k.

To see that this solves the problem, let $N = mn$ and number the squares of the $m \times n$ rectangle 1, …, N in some order. Let k be the number of ways of placing the figure on the rectangle, possibly after being rotated, so that the figure lies entirely over the rectangle, and number these placements 1, …, k. Let $\{e_j\}_{j=1}^N$ be the standard basis of $\mathbb{R}^N$. Then let $w = \sum_{j=1}^N e_j$ and define $w_i = \sum_{j=1}^N a_{ij} e_j$, where a_{ij} is 1 if the ith placement of the figure covers square j and 0 otherwise; the entries of w and w_i are obviously rational. Then the given condition says exactly that if $v \in \mathbb{R}^N$ and $w \cdot v > 0$ then $w_i \cdot v > 0$ for some i. Thus by the claim above there exist nonnegative rationals $\{c_i\}_{i=1}^k$ such that $w = \sum_{i=1}^k c_i w_i$. By putting these rationals over a common denominator we can find an integer $n > 0$ and nonnegative integers $\{n_i\}_{i=1}^k$ such that $nw = \sum_{i=1}^k n_i w_i$. Therefore if we put n_i copies of the figure in the ith placement, each square will be covered exactly n times. So it suffices to prove the claim above.

We will first show that it suffices to find nonnegative *reals* $\{c_i\}_{i=1}^k$ for which the conclusion holds. Let us call a vector all of whose entries are rational a *rational vector.*

Lemma 26. *Let $\{v_1, \dots, v_m\}$ be a collection of linearly independent rational vectors in $\mathbb{R}^n$. Then this set can be extended to a basis $\{v_1, \dots, v_n\}$ of rational vectors.*

Proof. Induct on $n - m$. If $n = m$ then $\{v_1, \dots, v_m\}$ is already a basis and we are done. Otherwise $\{v_1, \dots, v_m\}$ spans a subspace V of dimension smaller than n. Then $\mathbb{R}^n \setminus V$ is a nonempty open set; since the set of rational vectors is dense in $\mathbb{R}^n$, we can find a rational vector $v_{m+1} \in \mathbb{R}^n \setminus V$. Then $\{v_1, \dots, v_{m+1}\}$ is a linearly independent set of rational vectors and by the inductive hypothesis can be extended to a basis of rational vectors. □

Lemma 27. *Suppose $\{v_1, \dots, v_m\}$ is a linearly independent set of rational vectors in $\mathbb{R}^n$ and v is a rational vector with $v = \sum_{i=1}^m r_i v_i$ for some reals r_i. Then r_i is rational for each i.*

Proof. By the previous lemma we can extend $\{v_1, \ldots, v_m\}$ to a basis $\{v_1, \ldots, v_n\}$ of rational vectors. Then

$$v = \sum_{i=1}^{n} r_i v_i,$$

where we let $r_i = 0$ for $m < i \le n$. The r_i are then solutions of a nonsingular system of equations with rational coefficients; by Cramer's Rule, each r_i can be written as a rational function of the coefficients and is therefore rational. ☐

Lemma 28. *Suppose there exist nonnegative real numbers r_i such that $w = \sum_{i=1}^{k} r_i w_i$, for rational vectors w, w_i. Then there exist nonnegative rational numbers c_i such that $w = \sum_{i=1}^{k} c_i w_i$.*

Proof. We will "extend" w, w_1, …, w_k to vectors $\hat{w}$, $\hat{w}_1$, …, $\hat{w}_k$ in a larger space $\mathbb{R}^n$ such that $\hat{w} = \sum_{i=1}^{k} c_i \hat{w}_i$ for some reals $\{c_i\}_{i=1}^{k}$ and $\hat{w}_1$, …, $\hat{w}_k$ are linearly independent; the result will then follow from the previous lemma. Write $w = (w^1, \ldots, w^N)$, $w_i = (w_i^1, \ldots, w_i^N)$; we will choose

$$\begin{aligned}
\hat{w} &= (w^1, \ldots, w^N, x_1, x_2, \ldots, x_k), \\
\hat{w}_1 &= (w_1^1, \ldots, w_1^N, y_1, 0, \ldots, 0), \\
\hat{w}_2 &= (w_2^1, \ldots, w_2^N, 0, y_2, \ldots, 0), \\
&\vdots \\
\hat{w}_k &= (w_k^1, \ldots, w_k^N, 0, 0, \ldots, y_k)
\end{aligned}$$

where x_i is rational and $y_i \in \{0, 1\}$ for each i, so $\hat{w}$, $\hat{w}_1$, …, $\hat{w}_k$ will be rational vectors. I claim that this can be done in such a way that the system of equations

$$c_i \ge 0,\ \sum_{i=1}^{k} c_i w_i = w, \text{ and } c_i y_i = x_i \text{ for } 1 \le i \le k \qquad (1)$$

has a unique solution, and $y_i = 1$ if $c_i = 0$. We proceed by induction. Define

$T_j(c_1, \ldots, c_j) = \{ (d_1, \ldots, d_k) \in [0, \infty)^k \mid d_i = c_i$ for $1 \le i \le j$, $d_i y_i = x_i$ for $1 \le i \le j$, and $\sum_{i=1}^{k} d_i w_i = w \}$.

Suppose we have chosen x_i and y_i for $i < j$ such that there are unique reals c_1, …, c_{j-1} for which $T_{j-1}(c_1, \ldots, c_{j-1})$ is nonempty; call this set T. (If $j = 1$, then $T = T_0() = \{ (d_1, \ldots, d_k) \in [0, \infty)^k \mid \sum_{i=1}^{k} d_i w_i = w \}$, which is nonempty by assumption.) Then T is a nonempty convex subset of $\mathbb{R}^k$, so $S = \pi_j(T)$, the projection of T onto the jth axis, is a nonempty convex subset of $[0, \infty)$. If S contains more than one point, or if $S = \{0\}$, then S contains a nonnegative rational q; let $x_j = q$ and $y_j = 1$ and take $c_j = q$. Otherwise S contains a single positive real number r; let $x_r = y_r = 0$ and take $c_j = r$. I claim that $T_j(c'_1, \ldots, c'_j)$ is nonempty iff $(c'_1, \ldots, c'_j) = (c_1, \ldots, c_j)$. In either case $c_j \in S$ and $d_j = c_j$ implies $d_j y_j = x_j$, so

$$T_j(c_1, \ldots, c_j) = T_{j-1}(c_1, \ldots, c_{j-1}) \cap \pi_{j-1}(\{c_j\}) \neq \emptyset.$$

Conversely, suppose $T_j(c'_1, \ldots, c'_j)$ is nonempty. Then since $T_j(c'_1, \ldots, c'_j)$ is a subset of $T_{j-1}(c'_1, \ldots, c'_{j-1})$, $(c'_1, \ldots, c'_{j-1}) = (c_1, \ldots, c_{j-1})$ by the inductive hypothesis. Let $(d_1, \ldots, d_k) \in T_j(c_1, \ldots, c_{j-1}, c'_j) \subset T$; then in particular $c'_j = d_j \in \pi_j(T) = S$ and $d_j y_j = x_j$. If $y_j = 0$, then by construction $S = \{c_j\}$, so $c'_j = d_j = c_j$; otherwise $y_j = 1$ and $c'_j = d_j = x_j / y_j = x_j = c_j$. Hence in either case $c'_j = c_j$. Thus there are unique reals $(c_1, \ldots, c_j)$ such that $T_j(c_1, \ldots, c_j)$ is nonempty, as desired. Moreover, if $c_j = 0$, then $y_j = 1$ by construction. The claim follows by induction on j.

We now construct vectors $\hat{w}$, $\hat{w}_1$, $\ldots$, $\hat{w}_k$as described above. The previous result says that the equation $\hat{w} = \sum_{i=1}^k c_i \hat{w}_i$ has a unique solution $(c_1, \ldots, c_k)$ in $[0, \infty)^k$. We now prove that $\hat{w}_1$, $\ldots$, $\hat{w}_k$ are linearly independent. Suppose that $\sum_{i=1}^k s_i \hat{w}_i = 0$ for some reals s_i, not all 0; then $s_i y_i = 0$, so $s_i = 0$ if $y_i \neq 0$, in particular if $c_i = 0$. Therefore we can pick $\epsilon > 0$ so small that $|\epsilon s_i| \leq c_i$ for each i. But then

$$\hat{w} = \sum_{i=1}^k c_i \hat{w}_i + \sum_{i=1}^k \epsilon s_i \hat{w}_i = \sum_{i=1}^k (c_i + \epsilon s_i) \hat{w}_i$$

and $c_i + \epsilon s_i \geq 0$ for each i, which contradicts uniqueness. Hence the vectors $\hat{w}_1$, $\ldots$, $\hat{w}_k$ are linearly independent. Since $\hat{w}$, $\hat{w}_1$, $\ldots$, $\hat{w}_k$ are rational vectors and $\hat{w} = \sum_{i=1}^k c_i \hat{w}_i$, by the previous lemma c_i is rational for each i. Then by construction $c_i \geq 0$ and $w = \sum_{i=1}^k c_i w_i$, so we are done.

We will now find nonnegative real numbers r_1, $\ldots$, r_k such that $w = \sum_{i=1}^k r_i w_i$. Given $x = (x_1, x_2, \ldots, x_n) \in \mathbb{R}^n$, we define the *norm* of x by the equation

$$||x|| = \sqrt{x_1^2 + x_2^2 + \cdots + x_n^2}.$$

Let $S = \{\sum_{i=1}^{k} r_i w_i \mid r_i \geq 0\}$, and let $B = S \cap \{v \mid ||v|| \leq 2||w||\}$. Then B is a nonempty compact set, and the function $f(v) = ||w - v||$ is continuous, so there exists $v \in B$ such that $||w - v'|| \geq ||w - v||$ for all $v' \in B$. In particular, $||w - v|| \leq ||w - 0|| = ||w||$ as $0 \in B$, so if $v' \in S \setminus B$, $||w - v'|| \geq ||v'|| - ||w|| \geq ||w|| \geq ||w - v||$; thus in fact $||w - v'|| \geq ||w - v||$ for all $v' \in S$. If $||w - v|| = 0$, then $w \in S$ and we are done, so assume $||w - v|| > 0$; we will derive a contradiction.

First, we show that $w \cdot (w - v) > 0$. Note that $w \cdot (w - v) = (w - v) \cdot (w - v) + v \cdot (w - v) = ||w - v||^2 + v \cdot (w - v) > v \cdot (w - v)$, so it suffices to show $v \cdot (w - v) = 0$. Suppose $v \cdot (w - v) \neq 0$; then $v \neq 0$. Note that for $t \geq 0$, $tv \in S$ and

$$||w - tv||^2 = (w - tv) \cdot (w - tv) = w \cdot w - 2t(v \cdot w) + t^2(v \cdot v);$$

since this quadratic function attains its minimum at $t = 1$ by the choice of v, we must have $2(v \cdot w) = 2(v \cdot v)$, so $v \cdot (w - v) = 0$. Hence $w \cdot (w - v) > 0$ by the argument above.

We now show that $w_i \cdot (w - v) \leq 0$ for each i by a similar argument. Observe that $v + tw_i \in S$ for $t \geq 0$ and

$$||w - (v + tw_i)||^2 = (w - v) \cdot (w - v) - 2t(w_i \cdot (w - v)) + t^2(w_i \cdot w_i);$$

by construction this function attains its minimum over $[0, \infty)$ at 0, so we must have $w_i \cdot (w - v) \leq 0$. Hence we have arrived at a contradiction: $w \cdot (w - v) > 0$ but $w_i \cdot (w - v) \leq 0$ for each i.

Thus $||w - v|| = 0$, so $w \in S$ and $w = \sum_{i=1}^{k} r_i w_i$ for some nonnegative reals r_i; by the third lemma, we can find nonnegative rationals c_i for which $w = \sum_{i=1}^{k} c_i w_i$. By the argument at the beginning of the proof, this completes the problem.

1.14 Taiwan

Problem 1

Show that for positive integers m and n,

$$\gcd(m,n) = 2\sum_{k=0}^{m-1}\left\lfloor \frac{kn}{m}\right\rfloor + m + n - mn.$$

Solution. Let $g = \gcd(m,n), m = m'g$, and $n = n'g$. Then m' and n' are positive integers and $\gcd(m',n') = 1$. As k takes on values from 1 to $m-1$, $kn/m = kn'/m'$ is an integer if and only if $m'|k$, which happens

$$\left\lfloor \frac{m-1}{m'}\right\rfloor = \left\lfloor g - \frac{1}{m'}\right\rfloor = g - 1$$

times. We notice that

$$\left\lfloor \frac{kn}{m}\right\rfloor + \left\lfloor n - \frac{kn}{m}\right\rfloor = \begin{cases} n-1, & \text{if } \dfrac{kn}{m} \text{ is not an integer,} \\ n, & \text{if } \dfrac{kn}{m} \text{ is an integer.} \end{cases}$$

So

$$\sum_{k=1}^{m-1}\left(\left\lfloor \frac{kn}{m}\right\rfloor + \left\lfloor n - \frac{kn}{m}\right\rfloor\right) = (m-1)(n-1) + (g-1) = mn - m - n + g.$$

We have

$$\begin{aligned} 2\sum_{k=0}^{m-1}\left\lfloor \frac{kn}{m}\right\rfloor &= 2\sum_{k=1}^{m-1}\left\lfloor \frac{kn}{m}\right\rfloor \\ &= \sum_{k=1}^{m-1}\left\lfloor \frac{kn}{m}\right\rfloor + \sum_{k=1}^{m-1}\left\lfloor \frac{(m-k)n}{m}\right\rfloor \\ &= \sum_{k=1}^{m-1}\left(\left\lfloor \frac{kn}{m}\right\rfloor + \left\lfloor n - \frac{kn}{m}\right\rfloor\right) \\ &= mn - m - n + g \end{aligned}$$

and

$$\gcd(m,n) = g = 2\sum_{k=0}^{m-1}\left\lfloor \frac{kn}{m}\right\rfloor + m + n - mn$$

as desired.

Problem 2

Does there exist a solution to the equation

$$x^2 + y^2 + z^2 + u^2 + v^2 = xyzuv - 65$$

in integers x, y, z, u, v greater than 1998?

Solution. We say that a solution (x, y, z, u, v) to the equation is *good* if x, y, z, u, v are all positive integers and they are not all equal. We prove the following lemma.

Lemma 29. *If there is a good solution to the equation, then there are infinitely many good solutions to the equation.*

Proof. Let $(x_1, y_1, z_1, u_1, v_1)$ be a good solution. WLOG, let $x_1 \leq y_1 \leq z_1 \leq u_1 \leq v_1$. Since x_1, y_1, z_1, u_1, v_1 are not all equal, $x_1 < v_1$. The quadratic equation

$$x^2 - (y_1 z_1 u_1 v_1)x + (y_1^2 + z_1^2 + u_1^2 + v_1^2 + 65) = 0$$

has one integer solution $x = x_1$. The other solution is $x = x_2 = y_1 z_1 u_1 v_1 - x_1$, which is also an integer. We also notice that $x_1 x_2 = y_1^2 + z_1^2 + u_1^2 + v_1^2 + 65 > v_1^2$. Since $x_1 < v_1, x_2 > v_1$. We obtain a new good solution $(y_1, z_1, u_1, v_1, x_2)$. We can keep applying the procedure above to obtain infinitely many good solutions, as claimed. □

It is easy to check that $(1, 2, 3, 4, 5)$ and $(1, 1, 3, 8, 10)$ are good solutions. From the lemma, we can generate new good solutions and eventually make all of the terms greater than 1998.

Problem 3

Let m, n be positive integers, and let F be a collection of m-element subsets of $\{1, \ldots, n\}$ any two of which have nonempty intersection. Determine the maximum number of elements of F.

Solution. For $m > n/2$, we may take F to contain all the m-element subsets, so the maximum number of subsets is $\binom{n}{m}$. Thus we assume hereafter that $m \leq n/2$, in which case the maximum is $\binom{n-1}{m-1}$, achieved by taking all subsets containing a particular element. To establish the bound, note that for any arrangement of $\{1, \ldots, n\}$ around a circle, at most m intervals (sets of m consecutive numbers around the circle) can be in F. Indeed, suppose $\{a_1, \ldots, a_m\}$ is one such interval. Now each other interval appearing in F contains either $\{a_1, \ldots, a_j\}$ or $\{a_{j+1}, \ldots, a_m\}$

for some $j \in \{1, \ldots, m-1\}$. Moreover, for a given j, both types cannot occur. (Here we need $m \leq qn/2$, so the two sets don't "wrap around".) Thus at most m intervals appear in F.

By averaging over cyclic orderings, we see that at most m/n of the m-element subsets of $\{1, \ldots, n\}$ lie in F, proving the bound.

Problem 4

Let I be the incenter of triangle ABC, let D, E, F be the intersections of AI with BC, BI with CA, CI with AB, respectively, and let X, Y, Z be points on the lines EF, FD, DE. Show that

$$d(X, AB) + d(Y, BC) + d(Z, CA) \leq XY + YZ + ZX,$$

where $d(P, QR)$ denotes the distance from point P to line QR.

Solution. We begin with the equalities

$$d(X, BC) = d(X, CA) + d(X, AB)$$
$$d(Y, CA) = d(Y, AB) + d(Y, BC)$$
$$d(Z, AB) = d(Z, BC) + d(Z, CA).$$

Each of these holds by linearity. For example, the first equality holds for $X = E$ and $X = F$ by inspection, and X does not cross any of AB, BC, CA while moving along the segment EF, so the equality holds.

On the other hand,

$$XY \geq d(Y, CA) - d(X, CA)$$
$$YZ \geq d(Z, AB) - d(Y, AB)$$
$$ZX \geq d(X, BC) - d(Z, BC).$$

For example, the right side of the first inequality is the length of the component of XY perpendicular to CA, which cannot be longer than XY itself.

Putting the above equations and inequalities together yields the desired result.

Problem 5

For each positive integer n, let $w(n)$ be the number of (distinct) positive prime divisors of n. Find the smallest positive integer k such that for all

n,

$$2^{w(n)} \leq k\sqrt[4]{n}.$$

Solution. Let $p_1, p_2, p_3, \ldots$ denote the primes in increasing order. Then for n, $p_1 p_2 \cdots p_{j-1} \leq n < p_1 p_2 \cdots p_j$ for some positive integer j. Therefore, $w(n) \leq j - 1$ and

$$\frac{2^{w(n)}}{\sqrt[4]{n}} \leq \frac{2^{j-1}}{\sqrt[4]{p_1 p_2 \cdots p_{j-1}}}.$$

So

$$\begin{aligned} k &= \left\lceil \max\left(0, \frac{2^1}{\sqrt[4]{2 \cdot 3}}, \frac{2^2}{\sqrt[4]{2 \cdot 3 \cdot 5}}, \cdots\right) \right\rceil \\ &= \left\lceil \frac{2^6}{\sqrt[4]{2 \cdot 3 \cdot 5 \cdot 7 \cdot 11 \cdot 13}} \right\rceil = 5. \end{aligned}$$

It now suffices to prove that $k = 4$ does not work. This is easily shown by setting $n = 2 \cdot 3 \cdot 5 \cdot 7 = 210$.

1.15 Turkey

Problem 1

Squares $BAXX'$ and $CAYY'$ are erected externally on the sides of isosceles triangle ABC with equal sides $AB = AC$. Let E and F be the feet of the perpendiculars from an arbitrary point K on the segment BC to BY and CX, respectively. Let D be the midpoint of BC.

(a) Prove that $DE = DF$.

(b) Find the locus of the midpoint of EF.

Solution. Let BY and CX meet at P. A rotation by $\pi/2$ around A takes C onto Y and B onto X, so line CX is mapped to BY; thus $\angle BPC = \pi/2$. By symmetry, $BP = CP$, so BPC is an isosceles right triangle. It follows that $BE = EK = PF$, $\angle EBD = \pi/4 = \angle DPF$, and $BD = PD$, so triangles BDE and PDF are congruent; thus $DE = DF$. Finally, since $KEPF$ is a rectangle, the midpoint of EF is the midpoint of PK; as K varies along the segment BC, the locus of this point is the segment MN, where M is the midpoint of PB and N is the midpoint of PC.

Problem 2

Let $\{a_n\}$ be the sequence of real numbers defined by $a_1 = t$ and $a_{n+1} = 4a_n(1 - a_n)$ for $n \geq 1$. For how many distinct values of t do we have $a_{1998} = 0$?

First Solution. Let $f(x) = 4x(1 - x)$. Observe that

$$f^{-1}(0) = \{0, 1\}, \quad f^{-1}(1) = \{1/2\}, \quad f^{-1}([0, 1]) = [0, 1],$$

and $|\{y : f(y) = x\}| = 2$ for all $x \in [0, 1)$.

Let $A_n = \{x \in \mathbb{R} \mid f^n(x) = 0\}$; then

$$\begin{aligned} A_{n+1} &= \{x \in \mathbb{R} \mid f^{n+1}(x) = 0\} \\ &= \{x \in \mathbb{R} \mid f^n(f(x)) = 0\} = \{x \in \mathbb{R} \mid f(x) \in A_n\}. \end{aligned}$$

We claim that for all $n \geq 1$, $A_n \subset [0, 1]$, $1 \in A_n$, and $|A_n| = 2^{n-1} + 1$.

For $n = 1$, we have $A_1 = \{x \in \mathbb{R} \mid f(x) = 0\} = \{0, 1\}$, and the claims hold.

Now suppose $n \geq 1$ and $A_n \subset [0, 1]$, $1 \in A_n$, and $|A_n| = 2^{n-1} + 1$. Then $x \in A_{n+1} \Rightarrow f(x) \in A_n \subset [0, 1] \Rightarrow x \in [0, 1]$, so $A_{n+1} \subset [0, 1]$.

Since $f(0) = f(1) = 0$, we have $f^{n+1}(1) = 0$ for all $n \geq 1$, so $1 \in A_{n+1}$. Now we have

$$\begin{aligned}
|A_{n+1}| = |\{x : f(x) \in A_n\}| &= \sum_{a \in A_n} |\{x : f(x) = a\}| \\
&= |\{x : f(x) = 1\}| + \sum_{\substack{a \in A_n \\ a \in [0,1)}} |\{x : f(x) = a\}| \\
&= 1 + \sum_{\substack{a \in A_n \\ a \in [0,1)}} 2 = 1 + 2(|A_n| - 1) \\
&= 1 + 2(2^{n-1} + 1 - 1) \\
&= 2^n + 1
\end{aligned}$$

Thus the claim holds by induction.

Finally, $a_{1998} = 0$ iff $f^{1997}(t) = 0$ iff $t \in A_{1997}$, so there are $2^{1996} + 1$ such values of t.

Second Solution. As in the previous solution, observe that if $f(x) \in [0, 1]$ then $x \in [0, 1]$, so if $a_{1998} = 0$ we must have $t \in [0, 1]$. Now choose $\theta \in [0, \pi/2]$ such that $\sin\theta = \sqrt{t}$. Observe that for any $\phi \in \mathbb{R}$,

$$f(\sin^2\phi) = 4\sin^2\phi\,(1 - \sin^2\phi) = 4\sin^2\phi\cos^2\phi = \sin^2 2\phi;$$

since $a_1 = \sin^2\theta$, it follows that

$$a_2 = \sin^2 2\theta,\ a_3 = \sin^2 4\theta, \ldots, a_{1998} = \sin^2 2^{1997}\theta.$$

Therefore

$$a_{1998} = 0 \iff \sin 2^{1997}\theta = 0 \iff \theta = \frac{k\pi}{2^{1997}}$$

for some $k \in \mathbb{Z}$, so the values of t which give $a_{1998} = 0$ are $\sin^2(k\pi/2^{1997})$, $k \in \mathbb{Z}$, giving $2^{1996} + 1$ such values of t.

Problem 3

Let $A = \{1, 2, 3, 4, 5\}$. Find the number of functions f from the set of nonempty subsets of A to A for which $f(B) \in B$ for every $B \subseteq A$ and $f(B \cup C) \in \{f(B), f(C)\}$ for every B, $C \subseteq A$.

Solution. Consider the functions defined as follows: Choose some linear ordering $\prec$ on A, and for each $B \subseteq A$, let $f(B)$ be the smallest element of B according to the ordering $\prec$. These functions are all distinct, and clearly satisfy the given conditions. Thus it suffices to show that all functions satisfying the given conditions are of this form. Suppose f is a function satisfying these conditions, and suppose $B \subseteq C \subseteq A$ with $f(C) \in B$. Then $f(C) = f(B \cup (C \setminus B)) \in \{f(B), f(C \setminus B)\}$, but $f(C) \notin C \setminus B$ so $f(C \setminus B) \neq f(C)$; therefore $f(B) = f(C)$. Now let $a_1 = f(A)$; then for any $B \subseteq A$ containing a_1, $f(B) = a_1$ by the previous argument. Next let $a_2 = f(A \setminus \{a_1\})$; then for any $B \subseteq A$ such that $a_1 \notin B$ but $a_2 \in B$, we have $f(B) = a_2$. Similarly, let $a_3 = f(A \setminus \{a_1, a_2\})$; then if $a_1 \notin B$, $a_2 \notin B$, but $a_3 \in B$, we have $f(B) = a_3$. Define a_4 and a_5 similarly; then it follows that for each $B \subseteq A$, $f(B)$ is the smallest element of B according to the ordering $a_1 \prec a_2 \prec a_3 \prec a_4 \prec a_5$. Thus there are a total of $5! = 120$ such functions.

Problem 4

To n people are to be assigned n different houses. Each person ranks the houses in some order (with no ties). After the assignment is made, it is observed that every other assignment assigns at least one person to a house that person ranked lower than in the given assignment. Prove that at least one person received his/her top choice in the given assignment.

Solution. Label the n people P_1, P_2, …, P_n, and label their assigned houses H_1, H_2, …, H_n. Let $H_{f(i)}$ be the top choice of person P_i. Consider the sequence 1, $f(1)$, $f(f(1))$, …; since we only have finitely many houses, two of these must be equal, so we have a cycle i_1, $i_2 = f(i_1)$, …, $i_n = f(i_{n-1})$, $i_1 = f(i_n)$ for some $n \geq 1$. Now consider the assignment moving person P_{i_1} to house H_{i_2}, person P_{i_2} to house H_{i_3}, …, person P_{i_n} to house H_{i_1}; clearly this assignment assigns every person to a house that person ranked at least as high as in the given assignment. Thus by assumption this assignment must in fact be the original one. It follows that P_{i_1}, P_{i_2},…, P_{i_n} received their top choices in the given assignment.

Problem 5

Let ABC be a triangle. Suppose that the circle through C tangent to AB at A and the circle through B tangent to AC at A have different radii, and let D be their second intersection. Let E be the point on the ray AB such that $AB = BE$. Let F be the second intersection of the ray CA with the circle through A, D, E. Prove that $AF = AC$.

Solution. Invert around A. For point P, let P' denote the image of P under the inversion. Circles ABD and ACD go to lines $B'D'$ and $C'D'$ parallel to AC' and AB' respectively, so $AB'D'C'$ is a parallelogram. Since $AB = BE$, E' is the midpoint of AB', so triangles $AF'E'$ and $B'D'E'$ are congruent; in particular, $AF' = B'D' = AC'$ so $AF = AC$.

Problem 6

Let $f(x_1, \ldots, x_n)$ be a polynomial with integer coefficients of total degree less than n. Show that the number of ordered n-tuples $(x_1, \ldots, x_n)$ with $0 \le x_i \le 12$ such that $f(x_1, \ldots, x_n) \equiv 0 \pmod{13}$ is divisible by 13.

Solution. (All congruences in this problem are modulo 13.) We claim that

$$\sum_{x=0}^{12} x^k \equiv 0 \text{ for } 0 \le k < 12.$$

The case $k = 0$ is obvious, so suppose $k > 0$. Let g be a primitive root modulo 13; then the numbers g, $2g$, $\ldots$, $12g$ are 1, 2, $\ldots$, 12 in some order, so

$$\sum_{x=0}^{12} x^k \equiv \sum_{x=0}^{12} (gx)^k = g^k \sum_{x=0}^{12} x^k;$$

since $g^k \not\equiv 1$, we must have $\sum_{x=0}^{12} x^k \equiv 0$. This proves our claim.

Now let $S = \{(x_1, \ldots, x_n) \mid 0 \le x_i \le 12\}$. It suffices to show that the number of n-tuples $(x_1, \ldots, x_n) \in S$ with $f(x_1, \ldots, x_n) \not\equiv 0$ is divisible by 13, since $|S| = 13^n$ is divisible by 13. Consider the sum

$$\sum_{(x_1,\ldots,x_n)\in S} \left(f(x_1, \ldots, x_n)\right)^{12}.$$

This sum counts the number of n-tuples $(x_1, \ldots, x_n) \in S$ such that $f(x_1, \ldots, x_n) \not\equiv 0$, since by Fermat's Little Theorem

$$\left(f(x_1, \ldots, x_n)\right)^{12} \equiv \begin{cases} 1, & \text{if } f(x_1, \ldots, x_n) \not\equiv 0; \\ 0, & \text{if } f(x_1, \ldots, x_n) \equiv 0. \end{cases}$$

On the other hand, we can expand $\left(f(x_1, \ldots, x_n)\right)^{12}$ in the form

$$\left(f(x_1, \ldots, x_n)\right)^{12} = \sum_{j=1}^{N} c_j \prod_{i=1}^{n} x_i^{e_{ji}}$$

for some integers N, c_j, e_{ji}. Since f is a polynomial of total degree less than n, we have $e_{j1} + e_{j2} + \cdots + e_{jn} < 12n$ for every j, so for each j

there exists an i such that $e_{ji} < 12$. Thus by our claim

$$\sum_{(x_1,\ldots,x_n)\in S} c_j \prod_{i=1}^{n} x_i^{e_{ji}} = c_j \prod_{i=1}^{n} \sum_{x=0}^{12} x^{e_{ji}} \equiv 0$$

since one of the sums in the product is 0. Therefore

$$\sum_{(x_1,\ldots,x_n)\in S} \left(f(x_1,\ldots,x_n)\right)^{12} = \sum_{(x_1,\ldots,x_n)\in S} \sum_{j=1}^{N} c_j \prod_{i=1}^{n} x_i^{e_{ji}} \equiv 0,$$

so the number of $(x_1,\ldots,x_n)$ such that $f(x_1,\ldots,x_n) \not\equiv 0 \pmod{13}$ is divisible by 13 and we are done.

1.16 United Kingdom

Problem 1

A 5×5 square is divided into 25 unit squares. In each square is written one of 1, 2, 3, 4, 5 so that each row, column and diagonal contains each number once. The sum of the numbers in the squares immediately below the diagonal from top left to bottom right is called the score. Show that the score cannot equal 20, and find the maximum possible score.

Solution. Label the grid as a matrix with $(1,1)$ being the top left corner. Let $a_{i,j}$ be the number written in (i,j). A score is the sum $a_{2,1} + a_{3,2} + a_{4,3} + a_{5,4}$. Let D_1 denote the diagonal from $(0,0)$ to $(5,5)$; let D_2 denote the other diagonal; let a and b be two distinct values from 1, 2, 3, 4, 5. If $a_{2,1} = a_{4,3} = a$, then on D_1, $a_{5,5} = a$, and on D_2, $a_{1,5} = a$. So there are two a's on the 5th column, a contradiction. So $a_{2,1} \neq a_{4,3}$. Similarly, $a_{3,2} \neq a_{5,4}$. Thus there are at least 2 different values in each score. We consider the following cases.

(1) $(a_{2,1}, a_{3,2}, a_{4,3}, a_{5,4}) = (a, a, b, b)$. On D_2, $a_{1,5}$ equals to both a and b, which is impossible.

(2) $(a_{2,1}, a_{3,2}, a_{4,3}, a_{5,4}) = (a, b, b, a)$. Then $a_{3,3} = a$ and one of $a_{1,1}$ and $a_{5,5}$ takes value b. WLOG, let the $a_{1,1} = b$. On D_2, $a_{2,4} = b$. On the 5th column, $a_{5,5} = b$. Then there are two b's on D_1, contradiction.

From the above, we see that there are at least 3 different values in each score and the maximum of the score is less than or equal to $5+5+4+3 = 17$. The following example shows that it is indeed 17.

$$\begin{bmatrix} 1 & 5 & 4 & 3 & 2 \\ 5 & 3 & 2 & 4 & 1 \\ 2 & 4 & 5 & 1 & 3 \\ 4 & 1 & 3 & 2 & 5 \\ 3 & 2 & 1 & 5 & 4 \end{bmatrix}.$$

Problem 2

Let $a_1 = 19, a_2 = 98$. For $n \geq 1$, define a_{n+2} to be the remainder of $a_n + a_{n+1}$ when it is divided by 100. What is the remainder when

$$a_1^2 + a_2^2 + \cdots + a_{1998}^2$$

is divided by 8?

Solution. The answer is 0. Consider $a_n \pmod 4$ which is not changed by taking the remainder divided by 100, there's the cycle $3, 2, 1, 3, 0, 3$ which repeats 333 times. Then $a_1^2+a_2^2+\cdots+a_{1998}^2 \equiv 333(1+4+1+1+0+1) \equiv 0 \pmod 8$, as claimed.

Problem 3

Let ABP be an isosceles triangle with $AB = AP$ and $\angle PAB$ acute. Let PC be the line through P perpendicular to BP, with C a point on the same side of BP as A (and not lying on AB). Let D be the fourth vertex of parallelogram $ABCD$, and let PC meet DA at M. Prove that M is the midpoint of DA.

Solution. Let E be the point on BC such that $AE \parallel CP$. Then line AE is the perpendicular bisector of BP and $BE = EP$. Therefore, in right triangle CPB, PE is the median to the hypotenuse BC and $BE = EC$. Since $AECM$ is a parallelogram, $EC = AM$. Since $ABCD$ is a parallelogram, $AD = BC$. Thus $AM = EC = BC/2 = AD/2$, as desired.

Problem 4

Show that there is a unique sequence of positive integers (a_n) with $a_1 = 1$, $a_2 = 2$, $a_4 = 12$, and $a_{n+1}a_{n-1} = a_n^2 \pm 1$ for $n = 2, 3, 4, \ldots$.

Solution. First we claim that $\{a_n\}$ is a increasing sequence. We prove our claim by induction. For $n = 1$, We have $a_1 < a_2 < a_3 = 5 < a_4$. Suppose that $a_1 < \cdots < a_k$ for some $k \leq 4$, then $a_{k+1}a_{k-1} = a_k^2 \pm 1 \geq a_k(a_k - 1) \geq a_k a_{k-1}$ and $a_{k+1} > a_k$. Therefore the induction is complete. So $\{a_n\}$ is increasing. For $n > 3$, $a_{n-1} \geq 5$ and at most one of $a_n^2 + 1$ and $a_n^2 - 1$ is divisible by a_{n-1}. Therefore there are at most one such sequence.

Let $b_1 = 1$, $b_2 = 2$, $b_{n+2} = 2b_{n+1} + b_n$ for $n \geq 1$. Then $b_3 = 5$, $b_4 = 12$. We claim that $b_{n+1}b_{n-1} = b_n^2 + (-1)^n$ for $n = 2, 3, 4, \ldots$. We prove our claim by induction. For $n = 1, 2, 3$, it is evident. Suppose that $b_{k+1}b_{k-1} = b_k^2 + (-1)^k$ for some $k \geq 3$. We have

$$\begin{aligned} b_{k+2}b_k = b_{k+1}^2 + (-1)^{k+1} &\iff b_{k+2}b_k + b_{k+1}b_{k-1} = b_{k+1}^2 + b_k^2 \\ &\iff (b_{k+2} - b_k)b_k = (b_{k+1} - b_{k-1})b_{k+1} \\ &\iff 2b_{k+1}b_k = 2b_k b_{k+1}. \end{aligned}$$

So the induction is complete.

From the above we see that sequence $a_1 = 1, a_2 = 2,$

$$a_{n+2} = 2a_{n+1} + a_n, \qquad n = 1, 2, \ldots,$$

is the unique sequence of positive integers that satisfies the given conditions.

Problem 5

In triangle ABC, D is the midpoint of AB and E is the point of trisection of BC closer to C. Given that $\angle ADC = \angle BAE$, find $\angle BAC$.

First Solution. Let AE and CD meet at G; let F be the other trisection of BC. Then triangle ADG is isoceles and $AG = GD$. Since DF is the midline of triangle ABE, $DF \parallel AE$ and $DF \parallel GE$. So GE is the midline of triangle CDF and $DG = GC$. In triangle ADC, median AG is equal to half of the opposite side, so $\angle BAC = 90°$.

Second Solution. Extend AE to H such that $CH \parallel AD$. Then triangles CEH and BEA are similar and

$$CH = \frac{AB}{2} = AD.$$

Thus $ADHC$ is a parallelogram. Since $CH \parallel AD$,

$$\angle CHA = \angle EAD = \angle ADC$$

and $ADHC$ is cyclic. Therefore $ADHC$ is a rectangle and $\angle BAC = 90°$.

Problem 6

The ticket office at a train station sells tickets to 200 destinations. One day, 3800 passengers buy tickets. Show that at least 6 destinations receive the same number of passengers, and that there need not be 7 such destinations.

Solution. We prove the result by contradiction. Suppose, on the contrary, it's possible for at most 5 destinations having the same number of tickets. Then, at least $5(0 + 1 + 2 + \cdots + 39) = 3900$, which is absurd. For $k = 1, 2, \ldots, 33$, k tickets has been sold to exactly 6 destinations. Let the remaining destinations each have 217 tickets, making 3800 tickets. Thus, 7 is not necessary.

Problem 7

A triangle ABC has $\angle BAC > \angle BCA$. A line AP is drawn so that $\angle PAC = \angle BCA$, where P is inside the triangle. A point Q outside the triangle is constructed so that PQ is parallel to AB and BQ is parallel to AC. Let R be the point on BC (on the side of AP away from Q) such that $\angle PRQ = \angle BCA$. Prove that the circumcircles of ABC and PQR are tangent.

Solution. Let rays AP and BQ meet at S. Let ω_1 and ω_2 be the circumcircles of SPQ and SAB respectively. Since triangles SPQ and SAB are homothetic, ω_1 and ω_2 are tangent at S. From parallel rays,

$$\angle PSQ = \angle PAC = \angle ACB = \angle PRQ,$$

so ω_1 and ω_2 are the circumcircles of $PQSR$ and $ABSC$ respectively. Therefore circumcircles of ABC and PQR are tangent.

Problem 8

Let x, y, z be positive integers such that

$$\frac{1}{x} - \frac{1}{y} = \frac{1}{z}.$$

Let h be the greatest common divisor of x, y, z. Prove that $hxyz$ and $h(y-x)$ are perfect squares.

Solution. Let $x = ha$, $y = hb$, $z = hc$. Then a, b, c are positive integers such that $\gcd(a,b,c) = 1$. Let $\gcd(a,b) = g$. So $a = ga'$, $b = gb'$ and a' and b' are positive integers such that

$$\gcd(a',b') = \gcd(a'-b',b') = \gcd(a',a'-b') = 1.$$

We have

$$\frac{1}{a} - \frac{1}{b} = \frac{1}{c} \iff c(b-a) = ab \iff c(b'-a') = a'b'g.$$

So $g|c$ and $\gcd(a,b,c) = g = 1$. Therefore $\gcd(a,b) = 1$ and $\gcd(b-a,ab) = 1$. Thus $b - a = 1$ and $c = ab$. Now

$$hxyz = h^4abc = (h^2ab)^2 \quad \text{and} \quad h(y-x) = h^2$$

are both perfect squares, as desired.

Problem 9

Find all solutions of the system of equations

$$xy + yz + zx = 12$$

$$xyz = 2 + x + y + z$$

in positive real numbers x, y, z.

Solution. Let $\sqrt[3]{xyz} = a > 0$. From AM-GM,

$$12 = xy + yz + zx \geq 3a^2$$

$$a^3 = 2 + x + y + z \geq 2 + 3a.$$

Therefore $a^2 \leq 4$ and $a^3 - 3a - 2 = (a-2)(a-1)^2 \leq 0$. So all the equalities hold, $a = 2$ and $x = y = z$. Therefore $(x, y, z) = (2, 2, 2)$ is the only solution.

1.17 United States of America

Problem 1

Suppose that the set $\{1, 2, \cdots, 1998\}$ has been partitioned into disjoint pairs $\{a_i, b_i\}$ $(1 \leq i \leq 999)$ so that for all i, $|a_i - b_i|$ equals 1 or 6. Prove that the sum

$$|a_1 - b_1| + |a_2 - b_2| + \cdots + |a_{999} - b_{999}|$$

ends in the digit 9.

Solution. Let k denote the number of pairs $\{a_i, b_i\}$ with $|a_i - b_i| = 6$. Then the sum in question is $k \cdot 6 + (999 - k) \cdot 1 = 999 + 5k$, which ends in 9 provided k is even. Hence it suffices to show that k is even.

Write $k = k_{odd} + k_{even}$, where k_{odd} (resp. k_{even}) is equal to the number of pairs $\{a_i, b_i\}$ with a_i, b_i both odd (resp. even). Since there are as many even numbers as odd numbers between 1 and 1998, and since each pair $\{a_i, b_i\}$ with $|a_i - b_i| = 1$ contains one number of each type, we must have $k_{odd} = k_{even}$. Hence $k = k_{odd} + k_{even}$ is even as claimed.

Problem 2

Let C_1 and C_2 be concentric circles, with C_2 in the interior of C_1. From a point A on C_1 one draws the tangent AB to C_2 $(B \in C_2)$. Let C be the second point of intersection of AB and C_1, and let D be the midpoint of AB. A line passing through A intersects C_2 at E and F in such a way that the perpendicular bisectors of DE and CF intersect at a point M on AB. Find, with proof, the ratio AM/MC.

Solution. Writing the power of A with respect to C_2 we get $AE \cdot AF = AB^2$. On the other hand,

$$AD \cdot AC = \frac{AB}{2} \cdot 2AB = AB^2.$$

Hence $AE \cdot AF = AD \cdot AC$. This shows that triangles ADE and AFC (with the shared angle at A) are similar. Thus, $\angle AED = \angle ACF$, so $DEFC$ is cyclic. Since M is the intersection of the perpendicular bisectors of DE and CF, it must be the circumcenter of $DEFC$. Consequently, M also lies on the perpendicular bisector of CD. Since M is on AC, it must be the midpoint of CD. Hence $AM/MC = 5/3$.

Problem 3

Let $a_0, a_1, \ldots, a_n$ be numbers from the interval $(0, \pi/2)$ such that

$$\tan(a_0 - \frac{\pi}{4}) + \tan(a_1 - \frac{\pi}{4}) + \cdots + \tan(a_n - \frac{\pi}{4}) \geq n - 1.$$

Prove that

$$\tan a_0 \tan a_1 \cdots \tan a_n \geq n^{n+1}.$$

First Solution. Let $b_k = \tan(a_k - \pi/4)$, $k = 0, 1, \ldots, n$. It follows from the hypothesis that for each k, $-1 < b_k < 1$, and

$$1 + b_k \geq \sum_{0 \leq l \neq k \leq n} (1 - b_l). \tag{1}$$

Applying AM-GM to the positive numbers $1 - b_l$, $l = 0, 1, \ldots, k-1, k+1, \ldots, n$, we obtain

$$\sum_{0 \leq l \neq k \leq n} (1 - b_l) \geq n \left(\prod_{0 \leq l \neq k \leq n} (1 - b_l) \right)^{1/n}. \tag{2}$$

From (1) and (2) it follows that

$$\prod_{k=0}^{n} (1 + b_k) \geq n^{n+1} \left(\prod_{l=0}^{n} (1 - b_l)^n \right)^{1/n},$$

and hence that

$$\prod_{k=0}^{n} \frac{1 + b_k}{1 - b_k} \geq n^{n+1}.$$

Because

$$\frac{1 + b_k}{1 - b_k} = \frac{1 + \tan(a_k - \frac{\pi}{4})}{1 - \tan(a_k - \frac{\pi}{4})} = \tan\left(\left(a_k - \frac{\pi}{4}\right) + \frac{\pi}{4}\right) = \tan a_k,$$

the conclusion follows.

Second Solution. We first prove a short lemma.

Lemma 30. *Let w, x, y, z be real numbers with $x + y = w + z$ and $|x - y| < |w - z|$. Then $wz < xy$.*

Proof. Let $x + y = w + z = 2L$. Then there are non-negative numbers r, s with $r < s$ and

$$wz = (L - s)(L + s) < (L - r)(L + r) = xy,$$

as desired. □

We use the lemma in solving the problem. For $0 \leq k \leq n$, let $b_k = \tan(a_k - \pi/4)$ and let

$$t_k = \tan a_k = \frac{1 + b_k}{1 - b_k}.$$

Then $-1 < b_k < 1$ and

$$t_j t_k = \left(\frac{1+b_j}{1-b_j}\right)\left(\frac{1+b_k}{1-b_k}\right) = 1 + \frac{2}{\frac{1+b_j b_k}{b_j + b_k} - 1}. \tag{3}$$

First note that because $-1 < b_k < 1$ and $b_0 + b_1 + \cdots + b_n \geq n - 1$, it follows that $b_j + b_k > 0$ for all $0 \leq j,\ k \leq n$ with $j \neq k$. Next note that if $b_j + b_k > 0$ and $b_j \neq b_k$, then it follows from the lemma applied to (3) that the value of $t_j t_k$ can be made smaller by replacing b_j and b_k by two numbers closer together and with the same sum. In particular, if $b_j < 0$, then replacing b_j and b_k by their average reduces the problem to the case where $b_i > 0$ for all i.

We may now successively replace the b_i's by their arithmetic mean. As long as the b_i are not all equal, one is greater than the mean and another one is less than the mean. We can replace one of this pair by the arithmetic mean of all of the b_i's, and the other by a positive number chosen so that the sum of the pair does not change. Each such change decreases the product of the t_i's. It follows that for a given sum of the b_i's, the minimum product is attained when all of the b_i's are equal. In this case we have $b_i \geq \frac{n-1}{n+1}$, for each i, so

$$t_0 t_1 \cdots t_n \geq \left(\frac{1 + \frac{n-1}{n+1}}{1 - \frac{n-1}{n+1}}\right)^{n+1} = \left(\frac{2n}{2}\right)^{n+1} = n^{n+1}.$$

This completes the proof.

Third Solution. We present a solution based on calculus. Though all olympiad problems can be solved without calculus, solutions based on calculus are acceptable and may be instructive. We set

$$a = b_0 + b_1 + \cdots + b_n,$$

where $-1 < b_i < 1$, and assume that $a \geq n - 1$. We then show that the product

$$\prod_{k=0}^{n} \frac{1 + b_k}{1 - b_k}$$

attains its minimum when all of the b_k's are equal, that is, their common value is $a/(n+1)$. The desired inequality will follow immediately.

We proceed by induction. The case $n = 1$ was established in the discussion of (3) in the previous solution. For $n \geq 2$, set

$$\sum_{k=0}^{n-1} b_k = a' = a - b_n > n - 2.$$

The last inequality follows from $a \geq n-1$ and $b_n < 1$. Set $b = b_n$ and $c = a'/n$, so $b + nc = a$. By the induction hypothesis,

$$\left(\prod_{k=0}^{n-1} \frac{1+b_k}{1-b_k}\right) \frac{1+b_n}{1-b_n} \geq \left(\frac{1+c}{1-c}\right)^n \frac{1+b}{1-b}.$$

Thus we need to prove that

$$\left(\frac{1+c}{1-c}\right)^n \left(\frac{1+b}{1-b}\right) \geq \left(\frac{n+1+a}{n+1-a}\right)^{n+1}, \tag{4}$$

where the right hand side is obtained by substituting

$$b_k = \frac{a}{n+1}$$

for $k = 0, 1, \ldots, n$, in the product. Next, recall that a is fixed, and that $b + nc = a$. Thus we can eliminate b from (4) to obtain the equivalent inequality

$$\left(\frac{1+c}{1-c}\right)^n \left(\frac{1+a-nc}{1-a+nc}\right) \geq \left(\frac{n+1+a}{n+1-a}\right)^{n+1}. \tag{5}$$

Now bring all terms in (5) to the left side of the inequality, clear denominators, and replace c by x. Let the expression on the left define a function f with $f(x) = (1+x)^n(1+a-nx)(n+1-a)^{n+1} - (1-x)^n(1-a+nx)(n+1+a)^{n+1}$. To establish (5) it is sufficient to show that for $0 \leq x < 1$, $f(x)$ attains its minimum value at $x = a/(n+1)$. Towards this end we differentiate to obtain $f'(x) = n(a-(n+1)x)g(x)$, where

$$g(x) = (1+x)^{n-1}(n+1-a)^{n+1} - (1-x)^{n-1}(n+1+a)^{n+1}.$$

It is clear that $f'\left(\frac{a}{n+1}\right) = 0$, so we check the second derivative. We find

$$f''\left(\frac{a}{n+1}\right) = -n(n+1)g\left(\frac{a}{n+1}\right) > 0,$$

so f has a local minimum at $x = a/(n+1)$. But $f'(x)$ could have another zero, t, obtained by solving the equation $g(x) = 0$. Because

$$g'(x) = (n-1)(1+x)^{n-2}(n+1-a)^{n+1} + (n-1)(1-x)^{n-2}(n+1+a)^{n+1}$$

is obviously positive for all $x \in [0, 1)$, there is at most one solution to the equation $g(x) = 0$ in this interval. It is easy to check that $g(a/(n+1)) < 0$ and $g(1) > 0$. Thus there is a real number t, $a/(n + 1) < t < 1$, with $g(t) = 0$. For this t we have

$$f''(t) = n(a - (n + 1)t)g'(t) < 0.$$

Thus, t is a local maximum for f, and no other extrema exist on the interval $(0, 1)$.

The only thing left is to check that $f(1) \geq f(a/(n+1))$. Note that the case $x = 1$ is also an extreme case with $b_0 = b_1 = \cdots = b_{n-1} = 1$. This case does not arise in our problem, but we must check it to be sure that on the interval $0 \leq x < 1$, $f(x)$ has a minimum at $x = a/(n + 1)$. We have

$$f(1) = 2^n(1 + a - n)(n + 1 - a)^{n+1} \geq 0,$$

since $n - 1 \leq a \leq n + 1$, and $f(a/(n + 1)) = 0$ (by design). Thus $f(x)$ indeed attains a unique minimum at

$$x = \frac{a}{n+1},$$

as claimed.

Problem 4

A computer screen shows a 98×98 chessboard, colored in the usual way. One can select with a mouse any rectangle with sides on the lines of the chessboard and click the mouse button: as a result, the colors in the selected rectangle switch (black becomes white, white becomes black). Find, with proof, the minimum number of mouse clicks needed to make the chessboard all one color.

Solution. More generally, we show that the minimum number of selections required for an $n \times n$ chessboard is $n - 1$ if n is odd, and n if n is even. Consider the $4(n-1)$ squares along the perimeter of the chessboard, and at each step, let us count the number of pairs of adjacent perimeter squares which differ in color. This total begins at $4(n - 1)$, ends up at 0, and can decrease by no more than 4 each turn (If the rectangle touches two adjacent edges of the board, then only two pairs can be affected. Otherwise, the rectangle either touches no edges, one edge, or two opposite edges, in which case 0, 2 or 4 pairs change, respectively). Hence at least $n - 1$ selections are always necessary.

If n is odd, then indeed $n - 1$ selections suffice, by choosing every second, fourth, sixth, etc. row and column. However, if n is even, then $n - 1$ selections cannot suffice: at some point a corner square must be included in a rectangle (since the corners do not all begin having the same color), and such a rectangle can only decrease the above count by 2. Hence n selections are needed, and again by selecting every other row and column, we see that n selections also suffice.

Problem 5

Prove that for each $n \geq 2$, there is a set S of n integers such that $(a-b)^2$ divides ab for every distinct $a, b \in S$.

Solution. We will prove by induction on n, that we can find such a set S_n, all of whose elements are *nonnegative*. For $n = 2$, we may take $S_2 = \{0, 1\}$.

Now suppose that for some $n \geq 2$, the desired set S_n of n nonnegative integers exists. Let L be the least common multiple of those numbers $(a-b)^2$ and ab that are nonzero, with (a, b) ranging over pairs of distinct elements from S_n. Define

$$S_{n+1} = \{L + a \,:\, a \in S_n\} \cup \{0\}.$$

Then S_{n+1} consists of $n + 1$ nonnegative integers, since $L > 0$. If $\alpha, \beta \in S_{n+1}$ and either α of β is zero, then $(\alpha - \beta)^2$ divides $\alpha\beta$. If $L + a, L + b \in S_{n+1}$, with a, b distinct elements of S_n, then

$$(L + a)(L + b) \equiv ab \equiv 0 \pmod{(a - b)^2},$$

so $[(L + a) - (L + b)]^2$ divides $(L + a)(L + b)$, completing the inductive step.

Problem 6

Let $n \geq 5$ be an integer. Find the largest integer k (as a function of n) such that there exists a convex n-gon $A_1 A_2 \ldots A_n$ for which exactly k of the quadrilaterals $A_i A_{i+1} A_{i+2} A_{i+3}$ have an inscribed circle. (Here $A_{n+j} = A_j$.)

Solution. The maximum is $\lfloor n/2 \rfloor$. We first establish the upper bound by showing that if A, B, C, D, and E are consecutive vertices of the n-gon, then the quadrilaterals $ABCD$ and $BCDE$ cannot both have inscribed circles.

Assume the contrary. By equal tangents,

$$AB + CD = BC + AD$$

$$BC + DE = CD + BE$$

and so $AB+DE = AD+BE$. On the other hand, if O is the intersection of AD and BE, then by the triangle inequality, $AO + OB > AB$ and $OD + OE > DE$, so

$$AD + BE = AO + OB + OD + OE > AB + DE,$$

a contradiction. Now we give a construction to show that $\lfloor n/2 \rfloor$ circumscribing quadrilaterals are possible. First suppose that n is even. Draw an isosceles trapezoid with base angles $2\pi/n$ and admitting an inscribed circle. Let x be the length of the shorter base and y the length of either leg in the trapezoid. Then an equiangular n-gon with side lengths $x, y, x, y, \ldots$ clearly gives $n/2$ circumscribing quadrilaterals.

Now suppose that n is odd. Construct an $(n+1)$-gon $A_1 \ldots A_{n+1}$ yielding $(n+1)/2$ circumscribing quadrilaterals as described in the previous paragraph. Now erase A_{n+1} and move A_n to a new position so that the quadrilateral $A_{n-2}A_{n-1}A_nA_1$ has an inscribed circle; this gives $(n-1)/2$ circumscribing quadrilaterals, as desired.

1.18 Vietnam

Problem 1

Let $a \geq 1$ be a real number, and define the sequence $x_1, x_2, \ldots$ by $x_1 = a$ and

$$x_{n+1} = 1 + \log\left(\frac{x_n(x_n^2+3)}{3x_n^2+1}\right).$$

Prove that this sequence has a finite limit, and determine it.

Solution. We make the following observations for $x \geq 1$:

$$(x-1)^3 \geq 0 \Longrightarrow x^3 + 3x \geq 3x^2 + 1 \Longrightarrow \frac{x(x^2+3)}{3x^2+1} \geq 1$$

which implies that

$$1 + \log\frac{x(x^2+3)}{3x^2+1} \geq 1,$$

so inductively every term of the sequence is at least 1, and

$$1 \leq x^2 \Longrightarrow x^2 + 3 \leq 3x^2 + 1 \Longrightarrow \frac{x(x^2+3)}{3x^2+1} \leq x$$

which implies that

$$1 + \log\frac{x(x^2+3)}{3x^2+1} \leq 1 + \log x \leq x.$$

Thus the sequence is also monotonically decreasing; since it is bounded below, it must have a limit. The limit x must satisfy

$$x = 1 + \log\frac{x(x^2+3)}{3x^2+1}$$

and therefore the second inequality above must become an equality, which forces $x = 1$. So 1 is our limit.

Problem 2

Let P be a point on a given sphere. Three mutually perpendicular rays from P meet the sphere at A, B, C.

(a) Prove that for all such triples of rays, the plane of the triangle ABC passes through a fixed point.

(b) Find the maximum area of the triangle ABC.

First Solution. **(a)** We show that the point $1/3$ of the way from the center O to P must be the centroid of ABC and so lie on its plane. Let O be the center of a coordinate system and rotate it so that PA,PB,PC are parallel to the axes; then $P = (x_0, y_0, z_0)$ gives $A = (-x_0, y_0, z_0)$, $B = (x_0, -y_0, z_0)$, $C = (x_0, y_0, -z_0)$ and the centroid is

$$(x_0/3, y_0/3, z_0/3),$$

proving our claim.

(b) In the notation above, if R is the radius of the sphere, $AB^2 = (2x_0)^2 + (2y_0)^2$ and similarly for BC^2 and CA^2, so, adding gives

$$AB^2 + BC^2 + CA^2 = 8(x_0^2 + y_0^2 + z_0^2) = 8R^2.$$

Now let F be the Fermat point of ABC; we have $\angle AFB \geq 2\pi/3$. For fixed AB we find that the point F satisfying this constraint and farthest from AB is the one for which $\angle ABF = \angle BAF = \pi/6$, which gives $[ABF] = AB^2\sqrt{3}/12$, and so over all possible F we have $[ABF] \leq AB^2\sqrt{3}/12$. We have something similar for BCF and CAF; adding gives

$$[ABC] \leq \frac{(AB^2 + BC^2 + CA^2)\sqrt{3}}{12} = \frac{2\sqrt{3}R^2}{3}.$$

Equality is achieved when ABC is equilateral.

Second Solution. **(a)** Same as that of the first solution.

(b) First the $x_0 y_0 z_0 \neq 0$ or otherwise we obtain the degenerate case. Let $ax + by + cz = 1$ be an equation of the plane passing through A, B, C. We have

$$-ax_0 + by_0 + cz_0 = 1$$

$$ax_0 - by_0 + cz_0 = 1$$

$$ax_0 + by_0 - cz_0 = 1$$

We obtain that $a = 1/x_0$, $b = 1/y_0$, $c = 1/z_0$, and the plane passing through A, B, C satisfying the equation $x/x_0 + y/y_0 + z/z_0 - 1 = 0$. Thus the distance from P to the plane ABC is $d_{P-ABC} = 2/\sqrt{1/x_0^2 + 1/y_0^2 + 1/z_0^2}$. Now we calculate the volume of the tetrahedron $PABC$ in two ways and obtain

$$\frac{[ABC] \cdot d_{P-ABC}}{3} = V_{PABC} = \frac{PA \cdot PB \cdot PC}{6},$$

which implies that

$$\begin{aligned}
[ABC] &= 2\sqrt{x_0^2y_0^2z_0^2(1/x_0^2+1/y_0^2+1/z_0^2)}\\
&= 2\sqrt{x_0^2y_0^2+y_0^2z_0^2+z_0^2x_0^2}\\
&= \frac{2\sqrt{3(x_0^2y_0^2+y_0^2z_0^2+z_0^2x_0^2)}}{\sqrt{3}}\\
&\le \frac{2\sqrt{2(x_0^2y_0^2+y_0^2z_0^2+z_0^2x_0^2)+x_0^4+y_0^4+z_0^4}}{\sqrt{3}}\\
&= \frac{2(x_0^2+y_0^2+z_0^2)}{\sqrt{3}}\\
&= \frac{2\sqrt{3}R^2}{3}.
\end{aligned}$$

Equality holds when $PA = PB = PC$, which is equivalent to ABC is equilateral.

Problem 3

Let a, b be integers. Define a sequence $a_0, a_1, \ldots$ of integers by

$$a_0 = a, a_1 = b, a_2 = 2b - a + 2, a_{n+3} = 3a_{n+2} - 3a_{n+1} + a_n.$$

(a) Find the general term of the sequence.

(b) Determine all a, b for which a_n is a perfect square for all $n \ge 1998$.

Solution. **(a)** Answer: $a_n = n^2 + (b-a-1)n + a$. It is easy to check that any quadratic polynomial satisfies the given recurrence, so we need only find a quadratic which takes on the given values for $n = 0, 1, 2$, and it will provide the general term by induction. By Lagrange interpolation or otherwise we arrive at the answer.

(b) Answer: $(b-a-1)^2 = 4a$. We claim that a quadratic integer polynomial takes on square values for all sufficiently large n iff it is the square of a linear integer polynomial, which will prove our answer. One direction of the claim is obvious, so we prove the other. Suppose n^2+pn+q is a square for sufficiently large n, where p and q are integers. If p is odd then for sufficiently large n we have

$$(n + \frac{p-1}{2})^2 < n^2 + pn + q < (n + \frac{p+1}{2})^2,$$

contradiction; so p is even. Then for sufficiently large n we have

$$(n+\frac{p}{2}-1)^2 < n^2+pn+q < (n+\frac{p}{2}+1)^2$$

which implies that

$$n^2+pn+q = (n+\frac{p}{2})^2$$

and this is a polynomial identity, which proves the claim.

Problem 4

Let $x_1, \ldots, x_n$ $(n \geq 2)$ be positive numbers satisfying

$$\frac{1}{x_1+1998}+\frac{1}{x_2+1998}+\cdots+\frac{1}{x_n+1998}=\frac{1}{1998}.$$

Prove that

$$\frac{\sqrt[n]{x_1x_2\cdots x_n}}{n-1} \geq 1998.$$

Solution. Let

$$y_i = \frac{1998}{x_i+1998};$$

then $y_i \geq 0$ for each i and

$$y_1+\cdots+y_n = 1 \iff 1-y_i = \sum_{j\neq i} y_j,$$

and AM-GM gives

$$1-y_i \geq (n-1)\sqrt[n-1]{\prod_{j\neq i} y_j}.$$

Multiplying over all i gives

$$\prod_{i=1}^{n}(1-y_i) \geq (n-1)^n \prod_{i=1}^{n} y_i$$

which implies that

$$\prod_{i=1}^{n} \frac{1-y_i}{y_i} \geq (n-1)^n.$$

But

$$\frac{1-y_i}{y_i} = \frac{x_i}{1998},$$

so we obtain

$$x_1x_2\cdots x_n \geq 1998^n(n-1)^n$$

which is what we wanted.

Problem 5

Find the minimum of the expression

$$\sqrt{(x+1)^2+(y-1)^2}+\sqrt{(x-1)^2+(y+1)^2}+\sqrt{(x+2)^2+(y+2)^2}$$

over real numbers x, y.

First Solution. The answer is

$$2\sqrt{2}+\sqrt{6}.$$

We want to minimize the sum of the distances from the point $P=(x,y)$ to $A=(-1,1)$, $B=(1,-1)$, and $C=(-2,-2)$; the minimum is achieved when P is the Fermat point of triangle ABC, at which each side of the triangle subtends the angle $\frac{2\pi}{3}$. By symmetry, this point lies on the line $y=x$, and so if $D=(-1,-1)$ then $\angle PDB=\frac{\pi}{4}, \angle DPB=\frac{2\pi}{3}$ from which $\angle PBD=\frac{\pi}{12}$ so that $\frac{y+1}{1-y}=\tan\frac{\pi}{12}=2-\sqrt{3}\Rightarrow x=y=-\frac{\sqrt{3}}{3}$ which does indeed lie inside the triangle; now computation gives the desired answer.

Second Solution. To calculate the Fermat point P, we rotate A 60° counterclockwise around C to obtain B' and P is the intersection of the lines $y=x$ and BB'. We have

$$\overrightarrow{CB'}=\begin{bmatrix}\cos(\pi/3) & -\sin(\pi/3)\\ \sin(\pi/3) & \cos(\pi/3)\end{bmatrix}\cdot\overrightarrow{CA}$$

$$=\begin{bmatrix}\cos(\pi/3) & -\sin(\pi/3)\\ \sin(\pi/3) & \cos(\pi/3)\end{bmatrix}\cdot\begin{bmatrix}1\\3\end{bmatrix}=\begin{bmatrix}\frac{1-3\sqrt{3}}{2}\\ \frac{3+\sqrt{3}}{2}\end{bmatrix},$$

which implies that $B'=(-(5+3\sqrt{3})/2, (\sqrt{3}-1)/2)$. Routine computation leads to the desired result.

Problem 6

Prove that for each positive odd integer n, there is a unique polynomial $P(x)$ of degree n with real coefficients such that

$$P\left(x-\frac{1}{x}\right)=x^n-\frac{1}{x^n}$$

for $x\neq 0$. Also determine whether this assertion holds for n even.

Solution. Let P_k designate the desired polynomial of degree $2k+1$; we prove existence by strong induction. For $k=0$ we have $P_k(t)=t$. If $k>0$ then expanding by the binomial theorem and rearranging terms gives

$$\left(x-\frac{1}{x}\right)^{2k+1}=\sum_{j=0}^{k}(-1)^{k-j}\binom{2k+1}{k-j}\left(x^{2j+1}-\frac{1}{x^{2j+1}}\right),$$

and therefore we can take

$$P_k(t)=t^{2k+1}-\sum_{j=0}^{k-1}(-1)^{k-j}\binom{2k+1}{k-j}P_j(t).$$

Also, since the function $x-\frac{1}{x}$ takes on infinitely many values, if two P_k existed they would agree in infinitely many places and hence be identically equal; this proves uniqueness. Finally, no such polynomials can exist for n even: as

$$\lim_{x\to\infty} x^n-\frac{1}{x^n}=\infty \quad\text{and}\quad \lim_{x\to 0+} x^n-\frac{1}{x^n}=-\infty;$$

since a polynomial of even degree cannot take on both positive and negative values of arbitrarily large magnitude, our desired polynomial cannot exist.

2

1998 Regional Contests: Problems and Solutions

2.1 Asian Pacific Mathematics Olympiad

Problem 1

Let F be the set of all n-tuples $(A_1, \ldots, A_n)$ such that each A_i is a subset of $\{1, 2, \ldots, 1998\}$. Let $|A|$ denote the number of elements of the set A. Find

$$\sum_{(A_1,\ldots,A_n)\in F} |\bigcup_{i=1}^{n} A_i|.$$

Solution. This sum counts the total number of elements in the sets $\bigcup_{i=1}^{n} A_i$, $(A_1, \ldots, A_n) \in F$, which equals the total number of times an element of $\{1, 2, \ldots, 1998\}$ belongs to one of the sets $\bigcup_{i=1}^{n} A_i$. For any $x \in \{1, 2, \ldots, 1998\}$, there are $2^{1998n} - 2^{1997n}$ n-tuples $(A_1, \ldots, A_n) \in F$ such that $x \in \bigcup_{i=1}^{n} A_i$, since $x \notin \bigcup_{i=1}^{n} A_i$ if and only if $A_i \subset \{1, 2, \ldots, 1998\} \setminus \{x\}$ for each $i = 1, \ldots, n$, which happens for 2^{1997n} of the 2^{1998n} n-tuples in F. Thus

$$\begin{aligned}\sum_{(A_1,\ldots,A_n)\in F} |\bigcup_{i=1}^{n} A_i| &= \sum_{x\in\{1,2,\ldots,1998\}} 2^{1998n} - 2^{1997n} \\ &= 1998(2^{1998n} - 2^{1997n}).\end{aligned}$$

Problem 2

Show that for any positive integers a and b, $(36a + b)(a + 36b)$ cannot be a power of 2.

Solution. Write $a = 2^c p$, $b = 2^d q$, with p, q odd; assume without loss of generality that $c \geq d$. Then $36a + b = 36 \cdot 2^c p + 2^d q = 2^d(36 \cdot 2^{c-d} p + q)$, so $(36a + b)(36b + a) = 2^d(36 \cdot 2^{c-d} p + q)(36b + a)$ has the nontrivial odd factor $36 \cdot 2^{c-d} p + q$, and thus is not a power of 2.

Problem 3

Let a, b, c be positive real numbers. Prove that

$$\left(1 + \frac{a}{b}\right)\left(1 + \frac{b}{c}\right)\left(1 + \frac{c}{a}\right) \geq 2\left(1 + \frac{a+b+c}{\sqrt[3]{abc}}\right).$$

Solution. We have

$$\begin{aligned}
&\left(1 + \frac{a}{b}\right)\left(1 + \frac{b}{c}\right)\left(1 + \frac{c}{a}\right) \\
&\quad = 2 + \frac{a}{b} + \frac{b}{c} + \frac{c}{a} + \frac{a}{c} + \frac{b}{a} + \frac{c}{b} \\
&\quad = \left(\frac{a}{b} + \frac{a}{c} + \frac{a}{a}\right) + \left(\frac{b}{c} + \frac{b}{a} + \frac{b}{b}\right) + \left(\frac{c}{a} + \frac{c}{b} + \frac{c}{c}\right) - 1 \\
&\quad \geq 3\left(\frac{a}{\sqrt[3]{abc}} + \frac{b}{\sqrt[3]{abc}} + \frac{c}{\sqrt[3]{abc}}\right) - 1 \\
&\quad = 2\left(\frac{a+b+c}{\sqrt[3]{abc}}\right) + \left(\frac{a+b+c}{\sqrt[3]{abc}}\right) - 1 \\
&\quad \geq 2\left(\frac{a+b+c}{\sqrt[3]{abc}}\right) + 3 - 1 \\
&\quad = 2\left(1 + \frac{a+b+c}{\sqrt[3]{abc}}\right)
\end{aligned}$$

by two applications of AM-GM.

Problem 4

Let ABC be a triangle and D the foot of the altitude from A. Let E and F lie on a line passing through D such that AE is perpendicular to BE, AF is perpendicular to CF, and E and F are different from D. Let M and N be the midpoints of the segments BC and EF, respectively. Prove that AN is perpendicular to NM.

Solution. Since $\angle AEB = \angle ADB = \angle ADC = \angle AFC = \pi/2$, the quadrilaterals $ABDE$ and $ACDF$ are cyclic; thus $\angle EAB = \angle EDB =$

$\angle FDC = \angle FAC$. Also, $\angle AEB = \pi/2 = \angle AFC$, so triangles ABE and ACF are similar. Thus there exists a spiral similarity around A taking B to E and C to F. Since spiral similarities preserve midpoints, this transformation maps M to N, so triangles ABE and AMN are similar; thus $\angle ANM = \angle AEB = \pi/2$ and AN is perpendicular to NM.

Problem 5

Find the largest integer n such that n is divisible by all positive integers less than $\sqrt[3]{n}$.

Solution. The answer is 420, which satisfies the condition since $7 < \sqrt[3]{420} < 8$ and $420 = \operatorname{lcm}\{1,2,3,4,5,6,7\}$. Suppose $n > 420$ is an integer such that every positive integer less than $\sqrt[3]{n}$ divides n. Then $\sqrt[3]{n} > 7$, so $420 = \operatorname{lcm}\{1,2,3,4,5,6,7\}$ divides n; thus $n \geq 840$ and $\sqrt[3]{n} > 9$. Thus $2520 = \operatorname{lcm}\{1,2,\ldots,9\}$ divides n and $\sqrt[3]{n} > 13$. Now let m be the largest positive integer less than $\sqrt[3]{n}$, i.e., $m < \sqrt[3]{n} \leq m+1$. We have $m \geq 13$ and $\operatorname{lcm}\{1,2,\ldots,m\}$ divides n. But $\operatorname{lcm}\{m-3, m-2, m-1, m\}$ is at least $m(m-1)(m-2)(m-3)/6$ as 2 and 3 are the only possible common divisors of these four numbers. Thus

$$\frac{m(m-1)(m-2)(m-3)}{6} \leq n \leq (m+1)^3$$
$$\Longrightarrow m \leq 6\left(1+\frac{2}{m-1}\right)\left(1+\frac{3}{m-2}\right)\left(1+\frac{4}{m-3}\right).$$

The left-hand side of the inequality is an increasing function of m and the right hand side is a decreasing function of m. But for $m = 13$, we have

$$13 \cdot 12 \cdot 11 \cdot 10 = 17160 > 16464 = 6 \cdot 14^3,$$

so this inequality is false for all $m \geq 13$. Thus no $n > 420$ satisfies the given condition.

2.2 Austrian-Polish Mathematics Competition

Problem 1

Let x_1, x_2, y_1, y_2 be real numbers such that $x_1^2 + x_2^2 \le 1$. Prove the inequality

$$(x_1y_1 + x_2y_2 - 1)^2 \ge (x_1^2 + x_2^2 - 1)(y_1^2 + y_2^2 - 1).$$

Solution. If $y_1^2 + y_2^2 \ge 1$, then the LHS $\ge 0 \ge$ RHS and we are done. We suppose that $y_1^2 + y_2^2 < 1$. Let $a = 1 - x_1^2 - x_2^2$ and $b = 1 - y_1^2 - y_2^2$. Then $0 \le a, b \le 1$. We have

$$\begin{aligned}(x_1y_1 + x_2y_2 - 1)^2 &\ge (x_1^2 + x_2^2 - 1)(y_1^2 + y_2^2 - 1)\\ &\iff (2 - 2x_1y_1 - 2x_2y_2)^2 \ge 4ab\\ &\iff \left((x_1 - y_1)^2 + (x_2 - y_2)^2 + a + b\right)^2 \ge 4ab.\end{aligned}$$

But a, b are positive, so

$$\left((x_1 - y_1)^2 + (x_2 - y_2)^2 + a + b\right)^2 \ge (a+b)^2 \ge 4ab$$

as desired.

Problem 2

Consider n points $P_1, P_2, \ldots, P_n$ lying in that order on a straight line. We color each point in white, red, green, blue or violet. A coloring is admissible if for each two consecutive points P_i, P_{i+1} ($i = 1, 2, \ldots, n-1$) either both points are the same color, or at least one of them is white. How many admissible colorings are there?

Solution. We proceed inductively. Let us define the following three sequences: for each n, let a_n be the number of admissible colorings such that P_n is white, let b_n be the number of such colorings where P_n is not white, and let c_n be the total number of admissible colorings of n points. For $n > 2$, we obtain the following recurrence relations:

$$\begin{aligned}a_n &= c_{n-1}\\ b_n &= 4a_{n-1} + b_{n-1}\\ c_n &= a_n + b_n\end{aligned}$$

Thus, $b_n = b_{n-1} + 4c_{n-2}$ and $c_n = c_{n-1} + b_n$. Adding the two equations, we have $c_n = b_{n-1} + c_{n-1} + 4c_{n-2}$. Note that $b_n = c_n - c_{n-1}$, so $b_{n-1} = c_{n-1} - c_{n-2}$. Therefore

$$c_n = 2c_{n-1} + 3c_{n-2}.$$

Note that $c_1 = 5, c_2 = 13$; solving the characteristic equation $x^2 = 2x+3$, we have

$$c_n = \frac{3}{2} \cdot 3^n - \frac{1}{2} \cdot (-1)^n = \frac{3^{n+1} + (-1)^{n+1}}{2}.$$

Problem 3

Find all pairs of real numbers (x, y) satisfying the equations

$$2 - x^3 = y, \qquad 2 - y^3 = x.$$

Solution. If $x = y$, then $x^3 + x - 2 = (x-1)(x^2 + x + 2) = 0$, so $x = y = 1$.

Now we suppose that $x \neq y$. We claim that there is no solution. We prove our claim by contradiction. Suppose that there is a solution for the system. WLOG, say $x > y$. Then

$$2 - x^3 = y < x \iff 0 < x^3 + x - 2 = (x-1)(x^2 + x + 2),$$

so $x > 1$ and $x^3 > x$. Similarly, $y < 1$. We consider the following cases.

(a) $y \leq -1$ or $0 \leq y < 1$. Then $y \leq y^3$ and

$$2 = x^3 + y > x + y^3 = 2,$$

a contradiction;

(b) $-1 < y < 0$. Then $x = 2 - y^3 > 2$ and $x^3 > 8$. But then $y = 2 - x^3 < -6$, a contradiction.

Thus our assumption is false and there is no solution when $x \neq y$. So $x = y = 1$ is the only solution.

Problem 4

Let m, n be positive integers. Prove that

$$\sum_{k=1}^{n} \lfloor \sqrt[k^2]{k^m} \rfloor \leq n + m(2^{m/4} - 1).$$

Solution. First note that for all $k > m$, $\lfloor \sqrt[k^2]{k^m} \rfloor = 1$. This is because

$$k < 2^k \Longrightarrow k < 2^{\frac{k}{m} \cdot k} \iff k^m < 2^{k^2} \iff \sqrt[k^2]{k^m} < 2.$$

Thus, we will be done if we can show that

$$\sum_{k=1}^{m} \sqrt[k^2]{k^m} \leq m2^{m/4} = \sum_{k=1}^{m} 2^{m/4}.$$

But for $k = 1, 2, 3$ it is easy to check that

$$2^{k^2/4} \geq k \iff \sqrt[k^2]{k^m} \leq 2^{m/4}.$$

For $k \geq 4$, $2^{k^2/4} > 2^k > k$, so $\sqrt[k^2]{k^m} \leq 2^{m/4}$. So each term of LHS is greater than or equal to each term of RHS and we are done.

Problem 5

Find all pairs (a, b) of positive integers such that the equation

$$x^3 - 17x^2 + ax - b^2 = 0$$

has three integer roots (not necessarily distinct).

Solution. The solutions are

$$(a, b) = (80, 8) \text{ or } (80, 10) \text{ or } (90, 12) \text{ or } (88, 12).$$

Let

$$f(x) = x^3 - 17x^2 + ax - b^2,$$

and let r, s, t be its roots. Then $f(x) < 0$ for $x \leq 0$. Therefore, $r, s, t > 0$. Note that

$$17 = r + s + t, a = rs + st + tr, b^2 = rst.$$

It is not difficult to list the 24 possible sets $\{r, s, t\}$ such that $r+s+t = 17$ and find the ones that yield $rst = b^2$. These are:

$$\{r, s, t\} = \{1, 8, 8\} \text{ or } \{2, 5, 10\} \text{ or } \{3, 6, 8\} \text{ or } \{4, 4, 9\}$$

and they yield the (a, b) stated above.

Problem 6

Distinct points A, B, C, D, E, F lie on a circle in that order. The tangents to the circle at the points A and D, and the lines BF and CE, are concurrent. Prove that the lines AD, BC, EF are either parallel or concurrent.

First Solution. Let X be the point where the four lines meet and let O be the center of the circle.

We first consider the trivial case: $BC \parallel EF$. By symmetry, both BC and EF are perpendicular to OX. Since AD is also perpendicular to OX, we are done.

Now we consider the general case. We have the following lemma—a well known fact for people who are familar with polar maps.

Lemma 31. *Suppose S, T, U, V lie on circle ω. Then under the polar map with respect to ω, the polar of $TU \cap VS$ is the line through $UV \cap ST$ and $SU \cap TV$.*

Proof. The polar of $TU \cap VS$ is the line through the intersections of the polars of T, U and V, S. That is, it will be the line through the intersections of the tangents at T, U and the tangents at V, S. We can easily show by Pascal's Theorem that these two intersection points are collinear with $SU \cap VT$ and $UV \cap ST$: simply consider the inscribed hexagons $TTVUUS$ and $SSUVVT$. $\square$

Let $(S,T,U,V) = (B,C,E,F)$. Then from the lemma, $BC \cap EF$ is on AD, the polar of X and we are done.

Second Solution. Let line BC and line EF meet at Y. Connect YD and let YD meet ω at a second point A'. We claim that $A' = A$ and AD, BC, EF are concurrent at Y. Let G be the foot of the perpendicular from X to YD. Let P be a point on XY such that $BCXP$ is cyclic. From cyclic quadrilaterals $EFBC$ and $BCXP$, we have

$$\angle BPX = \angle BCE = \angle BFY,$$

so $YFBP$ is cyclic. From equal tangents and cyclic quadrilaterals, we have

$$\begin{array}{rcccl} XP \cdot XY &=& XA^2 &=& XD^2 \\ YP \cdot XY &=& YB \cdot YC &=& YA' \cdot YD. \end{array}$$

Adding the two equations yields

$$XY^2 = XD^2 + YA' \cdot YD \iff XY^2 - XD^2 = YA' \cdot YD.$$

Since $XY^2 = YG^2 + XG^2$ and $XD^2 = GD^2 + XG^2$,

$$\begin{aligned} XY^2 - XD^2 &= YG^2 - GD^2 \\ &= (YG + GD)(YG - GD) = YD \cdot (YG - GD). \end{aligned}$$

Thus

$$YA' \cdot YD = YD \cdot (YG - GD) \Rightarrow YG - GA' = YA' = YG - GD$$

and $GA' = GD$. So A' and D are symmetric about XG and XA' is tangent to ω at A', i.e., $A' = A$, as claimed.

Problem 7

Consider all pairs (a, b) of natural numbers such that the product $a^a b^b$, written in base 10, ends with exactly 98 zeroes. Find the pair (a, b) for which the product ab is smallest.

Solution. Let a_2 be the maximum integer such that $2^{a_2}|a$. Define a_5, b_2, and b_5 similarly. Our task translates into the following: find a, b such that $\min\{a_5 a + b_5 b, a_2 a + b_2 b\} = 98$ and ab is minimal. Since $5|a_5 a + b_5 b$, $a_5 a + b_5 b > 98$ and $\min\{a_5 a + b_5 b, a_2 a + b_2 b\} = a_2 a + b_2 b = 98$. Note that if $5|\gcd(a, b)$, then $a_2 a + b_2 b \neq 98$, contradiction. WLOG, suppose that $a_5 \geq 1$ and $b_5 = 0$. Let $a = 2^{a_2} 5^{a_5} x$ and $b = 2^{b_2} y$. ($\gcd(2, x) = \gcd(5, x) = \gcd(2, y) = 1$.) Then $a_5 a = a_5(2^{a_2} 5^{a_5} x) > 98$ and $a_2 a = a_2(2^{a_2} 5^{a_5} x) \leq 98$. So $a_5 > a_2$. We consider the following cases.

(a) $a_2 = 0$. Then $b_2(2^{b_2} y) = 98$. So $b_2 = 1, y = 49, b = 98$. Since $a_5(5^{a_5} x) \geq 98$ and x is odd $a = 5^{a_5} x \geq 125$ for $a_5 \geq 3$; $x \geq 3$ and $a \geq 75$ for $a_5 = 2$; $x \geq 21$ and $a \geq 105$ for $a_5 = 1$. Hence for $a_2 = 0$, $b = 98, a \geq 75$.

(b) $a_2 \geq 1$. Then $a_5 \geq 2$. We have $2^{a_2} 5^{a_5} x \leq 98$ and $5^{a_5} x \leq 49$. Thus $a_5 = 2, x = 1, a_2 = 1, a = 50$. Then $b_2 b = 48$. Let $b = 2^{b_2} y$. Then $b_2(2^{b_2} y) = 48$, which is impossible.

From the above, we have $(a, b) = (75, 98)$ or $(98, 75)$.

Problem 8

Let $n > 2$ be a given natural number. In each unit square of an infinite grid is written a natural number. A polygon is admissible if it has area n and its sides lie on the grid lines. The sum of the numbers written in the squares contained in an admissible polygon is called the value of the polygon. Prove that if the values of any two congruent admissible polygons are equal, then all of the numbers written in the squares of the grid are equal.

Solution. We superimpose the grid onto the coordinate plane such that the axes lie on grid lines. We consider two cases:

(a) n is odd. Let s_1 and s_2 be two adjacent unit squares, and let e be their common edge. Let T be an $n-1$ by 1 rectangle adjacent to both s_1 and s_2 and symmetric about e. Then $s_1 \cup T$ and $s_2 \cup T$ are two congruent admissible polygons, so if they have the same value, then the values in s_1 and s_2 are the same. It follows that all of the numbers written in the squares are equal.

(b) n is even. Let s_1 and s_2 be two squares separated by exactly one square s. Suppose that s_1 and s share the common edge a; s_2 and s share the common edge b; and that $a \parallel b$. Let T be an $n-1$ by 1 rectangle adjacent to both s_1 and s_2 and symmetric with respect to s. Then $s_1 \cup T$ and $s_2 \cup T$ are two congruent admissible polygons, so if they have the same value, then the values in s_1 and s_2 are the same. Thus, if we color the grid like a chess board, all the white squares share a common value x and all the black squares share a common value y.

Now consider the following two admissible polygons. The first is defined by a row of $n-2$ unit squares with one unit square added to each side, at one end. (This looks like a T-shape, with a very short top-of-T and a very long bottom-of-T.) The second is obtained by translating the first polygon by the vector $[0,1]$. One of the polygons has value $(n-2)(x+y)/2+2x$, and the other has value $(n-2)(x+y)/2+2y$. Thus $x=y$ and we are done.

Problem 9

Let K, L, M be the midpoints of sides BC, CA, AB, respectively, of triangle ABC. The points A, B, C divide the circumcircle of ABC to three arcs AB, BC, CA. Let X be the midpoint of the arc BC not containing A, let Y be the midpoint of the arc CA not containing B, and let Z be the midpoint of the arc AB not containing C. Let R be the circumradius and r the inradius of ABC. Prove that

$$r + KX + LY + MZ = 2R.$$

Solution. Let O be the center of the circumcircle. Notice that KX is the perpendicular bisector of BC. So O is on KX and $KX = OX - OK = R - OX$. Similarly, $LY = R - OL$, $MZ = R - OM$ and we are left to prove that $OM + OK + OL = R + r$. But

$$OM = R\cos\angle C, OK = R\cos\angle A, OL = R\cos\angle B.$$

So we need to prove that

$$\cos\angle A+\cos\angle B+\cos\angle C=1+\frac{r}{R}.$$

This is actually a well known fact, but we shall show this here anyway. We have

$$\begin{aligned}4R^2\sin\angle A\sin\angle B\sin\angle C &= AB\cdot AC\sin\angle A\\ &= 2[ABC]\\ &= r(AB+BC+CA)\\ &= 2rR(\sin\angle A+\sin\angle B+\sin\angle C).\end{aligned}$$

So

$$2R\sin\angle A\sin\angle B\sin\angle C=r(\sin\angle A+\sin\angle B+\sin\angle C).$$

But $\sin 2x=2\sin x\cos x$ gives

$$\begin{aligned}&\sin\angle A+\sin\angle B+\sin\angle C\\ &\quad=2\sin\frac{\angle A+\angle B}{2}\cos\frac{\angle A-\angle B}{2}+2\sin\frac{\angle C}{2}\cos\frac{\angle C}{2}\\ &\quad=2\cos\frac{\angle C}{2}\left(\cos\frac{\angle A-\angle B}{2}+\sin\frac{\angle C}{2}\right)\\ &\quad=2\cos\frac{\angle C}{2}\left(\cos\frac{\angle A-\angle B}{2}+\cos\frac{\angle A+\angle B}{2}\right)\\ &\quad=4\cos\frac{\angle A}{2}\cos\frac{\angle B}{2}\cos\frac{\angle C}{2}.\end{aligned}$$

So

$$\frac{r}{R}=4\sin\frac{\angle A}{2}\sin\frac{\angle B}{2}\sin\frac{\angle C}{2}.$$

We have

$$\begin{aligned}&\cos\angle A+\cos\angle B+\cos\angle C-1\\ &\quad=2\cos\frac{\angle A+\angle B}{2}\cos\frac{\angle A-\angle B}{2}+\cos\angle C-1\\ &\quad=2\sin\frac{\angle C}{2}\cos\frac{\angle A-\angle B}{2}-2\sin^2\frac{\angle C}{2}\\ &\quad=2\sin\frac{\angle C}{2}\left(\cos\frac{\angle A-\angle B}{2}-\sin\frac{\angle C}{2}\right)\end{aligned}$$

$$= 2\sin\frac{\angle C}{2}\left(\cos\frac{\angle A - \angle B}{2} - \cos\frac{\angle A + \angle B}{2}\right)$$

$$= 4\sin\frac{\angle C}{2}\sin\frac{\angle A}{2}\sin\frac{\angle B}{2}.$$

Thus $\cos\angle A + \cos\angle B + \cos\angle C = 1 + r/R$, as desired.

2.3 Balkan Mathematical Olympiad

Problem 1

Find the number of different terms of the finite sequence $\left\lfloor \frac{k^2}{1998} \right\rfloor$, where $k = 1, 2, \ldots, 1997$ and $\lfloor x \rfloor$ denotes the integer part of x.

Solution. Note that

$$\left\lfloor \frac{998^2}{1998} \right\rfloor = 498 < 499 = \left\lfloor \frac{999^2}{1998} \right\rfloor,$$

so we can compute the total number of distinct terms by considering $k = 1, \ldots, 998$ and $k = 999, \ldots, 1997$ independently. Observe that for $k = 1, \ldots, 997$,

$$\frac{(k+1)^2}{1998} - \frac{k^2}{1998} = \frac{2k+1}{1998} < 1,$$

so for $k = 1, \ldots, 998$, each of the numbers

$$\left\lfloor \frac{1^2}{1998} \right\rfloor = 0, 1, \ldots, 498 = \left\lfloor \frac{998^2}{1998} \right\rfloor$$

appears at least once in the sequence $\lfloor k^2/1998 \rfloor$ for a total of 499 distinct terms. For $k = 999, \ldots, 1996$, we have

$$\frac{(k+1)^2}{1998} - \frac{k^2}{1998} = \frac{2k+1}{1998} > 1,$$

so the numbers $\lfloor k^2/1998 \rfloor$ $(k = 999, \ldots, 1997)$ are all distinct, giving $1997 - 999 + 1 = 999$ more terms. Thus the total number of distinct terms is 1498.

Problem 2

If $n \geq 2$ is an integer and $0 < a_1 < a_2 \cdots < a_{2n+1}$ are real numbers, prove the inequality:

$$\sqrt[n]{a_1} - \sqrt[n]{a_2} + \sqrt[n]{a_3} - \cdots - \sqrt[n]{a_{2n}} + \sqrt[n]{a_{2n+1}} < \sqrt[n]{a_1 - a_2 + a_3 - \cdots - a_{2n} + a_{2n+1}}.$$

Solution. We will prove a slightly stronger statement by induction:
For any $n \geq 1$, $m \geq 2$, $0 < a_1 < a_2 < \cdots < a_{2n+1}$, we have

$$\sqrt[m]{a_1} - \sqrt[m]{a_2} + \sqrt[m]{a_3} - \cdots - \sqrt[m]{a_{2n}} + \sqrt[m]{a_{2n+1}} < \sqrt[m]{a_1 - a_2 + a_3 - \cdots - a_{2n} + a_{2n+1}}.$$

First, suppose $n = 1$. Define

$$f(t) = \sqrt[m]{t - a_2 + a_3} - \sqrt[m]{t}$$

for $t > 0$; then

$$f'(t) = \frac{1}{m}\left((t - a_2 + a_3)^{1/m-1} - t^{1/m-1}\right) < 0$$

for all $t > 0$, so f is strictly decreasing. Therefore

$$\sqrt[m]{a_3} - \sqrt[m]{a_2} = f(a_2) < f(a_1) = \sqrt[m]{a_1 - a_2 + a_3} - \sqrt[m]{a_1},$$

giving the desired result.

Now, suppose $n \geq 2$ and the statement holds for all smaller values of n. Then

$$\sqrt[m]{a_1} - \sqrt[m]{a_2} + \sqrt[m]{a_3} - \cdots - \sqrt[m]{a_{2n-2}} + \sqrt[m]{a_{2n-1}}$$
$$< \sqrt[m]{a_1 - a_2 + a_3 - \cdots - a_{2n-2} + a_{2n-1}},$$

and $a_1 < a_2$, $a_3 < a_4$, $\ldots$, so

$$(a_1 - a_2) + (a_3 - a_4) \cdots (a_{2n-1} - a_{2n-2}) + a_{2n-1}$$
$$\leq a_{2n-1} < a_{2n} < a_{2n+1};$$

thus

$$\sqrt[m]{a_1} - \sqrt[m]{a_2} + \sqrt[m]{a_3} - \cdots - \sqrt[m]{a_{2n}} + \sqrt[m]{a_{2n+1}}$$
$$< \sqrt[m]{a_1 - a_2 + a_3 - \cdots - a_{2n-2} + a_{2n-1}} - \sqrt[m]{a_{2n}} + \sqrt[m]{a_{2n+1}}$$
$$< \sqrt[m]{a_1 - a_2 + a_3 - \cdots - a_{2n-2} + a_{2n-1} - a_{2n} + a_{2n+1}}$$

by the case $n = 1$, and the induction is complete. The case $n = m \geq 2$ gives the statement of the problem.

Problem 3

Let T be a point inside triangle ABC, and let S be the set of all points except T that are inside or on the border of triangle ABC. Prove that S can be represented as a union of closed segments no two of which have a point in common. (A closed segment contains both of its ends.)

Solution. Suppose $WXYZ$ is a trapezoid with $WX \parallel YZ$. Let $[WXYZ] - [WX]$ denote the boundary and interior of $WXYZ$, minus the closed segment WX. Then we can decompose $[WXYZ] - [WX]$ into a union of disjoint closed segments parallel to WX, one joining each point on the half-open segment $(WZ]$ to the corresponding point on $(XY]$. Now construct $D \in (BC)$, $E \in (CA)$, $F \in (AB)$ such that $TD \parallel AB$, $TE \parallel BC$, and $TF \parallel CA$. Decomposing each of the regions $[TDBF]-[TD]$, $[TECD]-[TE]$, and $[TFAE]-[TF]$ in this way gives a representation of S as a disjoint union of closed segments.

Problem 4

Prove that the equation $y^2 = x^5 - 4$ has no integer solutions.

Solution. We consider the equation mod 11. Since

$$(x^5)^2 = x^{10} \equiv 0 \text{ or } 1 \pmod{11}$$

for all x, we have $x^5 \equiv -1, 0, \text{ or } 1 \pmod{11}$, so the right-hand side is either 6, 7, or 8 modulo 11. However, all squares are 0, 1, 3, 4, 5, or 9 modulo 11, so the equation $y^2 = x^5 - 4$ has no integer solutions.

2.4 Czech-Slovak Match

Problem 1

A polynomial $P(x)$ of degree $n \geq 5$ with integer coefficients and n distinct integer roots is given. Find all integer roots of $P(P(x))$ given that 0 is a root of $P(x)$.

Solution. The roots of $P(x)$ are clearly integer roots of $P(P(x))$; we claim there are no other integer roots. We prove our claim by contradiction. Suppose, on the contrary, that $P(P(k)) = 0$ for some integer k such that $P(k) \neq 0$. Let

$$P(x) = a(x - r_1)(x - r_2)(x - r_3) \cdots (x - r_n),$$

where $a, r_1, r_2, \ldots, r_n$ are integers,

$$r_1 = 0 \leq |r_2| \leq |r_3| \leq \cdots \leq |r_n|.$$

Since $P(k) \neq 0$, we must have $|k - r_i| \geq 1$ for all i. And since the r_i are all distinct, at most two of $|k - r_2|, |k - r_3|, |k - r_4|$ equal 1, thus

$$|a(k - r_2) \cdots (k - r_{n-1})| \geq |a||k - r_2||k - r_3||k - r_4| \geq 2.$$

So, $|P(k)| \geq 2|k(k - r_n)|$. Also note that $P(k) = r_{i_0}$ for some i_0 so $|P(k)| \leq |r_n|$. Now we consider the following two cases:

(a) $|k| \geq |r_n|$. Then $|P(k)| \geq 2|k(k - r_n)| \geq 2|k| > |r_n|$, a contradiction.

(b) $|k| < |r_n|$, i.e., $1 \leq |k| \leq |r_n| - 1$. Let a, b, c be real numbers, $a \leq b$. For $x \in [a, b]$, the funcion

$$f(x) = x(c - x)$$

reaches its minimum value at an endpoint $x = a$ or $x = b$, or at both endpoints. Thus

$$\begin{aligned}|k(k - r_n)| &= |k||r_n - k| \geq |k|(|r_n| - |k|) \geq |r_n| - 1\\ &\Longrightarrow |r_n| \geq |P(k)| \geq 2|k(k - r_n)| \geq 2(|r_n| - 1)\\ &\Longrightarrow |r_n| \leq 2.\end{aligned}$$

Since $n \geq 5$, this is only possible if

$$P(x) = (x + 2)(x + 1)x(x - 1)(x - 2).$$

But then it impossible to have $k \neq r_i$ and $|k| \leq |r_n|$, a contradiction.

Thus our assumption was incorrect, and the integer roots of $P(P(x))$ are exactly all the integer roots of $P(x)$.

Problem 2

The lengths of the sides of a convex hexagon $ABCDEF$ satisfy $AB = BC$, $CD = DE$, $EF = FA$. Show that

$$\frac{BC}{BE} + \frac{DE}{DA} + \frac{FA}{FC} \geq \frac{3}{2}.$$

Solution. Let $a = CE$, $c = EA$, $e = AC$. Applying Ptolemy's Inequality to quadrilateral $ABCE$, we have

$$\begin{aligned} BC(a+c) &= BC(CE + AE) \\ &= AB \cdot CE + BC \cdot AE \geq AC \cdot BE = BE \cdot e, \end{aligned}$$

which implies that

$$\frac{BC}{BE} \geq \frac{e}{a+c}.$$

Similarly,

$$\frac{DE}{DA} \geq \frac{a}{c+e}, \quad \frac{FA}{FC} \geq \frac{c}{e+a}.$$

Now by AM-HM, we have

$$\begin{aligned} \frac{BC}{BE} + \frac{DE}{DA} + \frac{FA}{FC} &\geq \frac{e}{a+c} + \frac{a}{c+e} + \frac{c}{e+a} \\ &= (a+c+e)\left(\frac{1}{a+c} + \frac{1}{c+e} + \frac{1}{e+a}\right) - 3 \\ &\geq \frac{9(a+c+e)}{(a+c)+(c+e)+(e+a)} - 3 = \frac{3}{2}. \end{aligned}$$

Problem 2

Find all functions $f : \mathbb{N} \to \mathbb{N} - \{1\}$ such that for all $n \in \mathbb{N}$,

$$f(n) + f(n+1) = f(n+2)f(n+3) - 168.$$

Solution. We claim that, for $n = 1, 2, \ldots$,

$$(f(2n), f(2n+1)) = (a, b) = (14, 14) \text{ or } (2, 170) \text{ or } (170, 2).$$

Since $f(n) \in \mathbb{N} - \{1\}$, $f(n) \geq 2$ and

$$(f(n+2) - 2)(f(n+3) - 2) \geq 0$$

$$2(f(n+2)+f(n+3)) \leq f(n+2)f(n+3)+4$$
$$2(f(n+2)+f(n+3)) \leq f(n)+f(n+1)+172.$$

Thus, eventually $f(n)+f(n+1) \leq 172$ for all n greater than some value N so that $f(n) < 172$ for all $n > N$. There are only finitely many triples of natural numbers less than 172, so eventually some 3-tuple of consecutive terms repeats. But since any 3 consecutive terms determine all other values of $f(i)$, the sequence must be periodic. So we can find $x \geq 3$ such that $f(x) = M = \max\{f(n)\}$. Then

$$f(x-2)+f(x-1) = Mf(x+1)-168$$
$$f(x-1)+M = f(x+1)f(x+2)-168,$$

which implies that

$$0 \leq M - f(x-2) = f(x+1)(f(x+2)-M) \leq 0.$$

So $f(x-2) = f(x+2) = M$. Thus every other term of the sequence $\{f(n)\}$ equals the maximum value M. Similarly, we can prove that every other term of the sequence $\{f(n)\}$ equals the minimum value m. Thus $\{f(n)\}$ equals $a, b, a, b, \ldots$ with $a+b = ab-168 \iff (a-1)(b-1) = 169$ and the presented solutions follow.

Problem 4

At a summer camp there are n girls $D_1, D_2, D_3, \ldots, D_n$ and $2n-1$ boys $C_1, C_2, C_3, \ldots, C_{2n-1}$. Girl D_i knows boys $C_1, \ldots, C_{2i-1}$ and no others. Prove that the number of ways to choose r boy-girl pairs so that each girl in a pair knows the boy in the pair is

$$\binom{n}{r}\frac{n!}{(n-r)!}.$$

Solution. Let $f(n,r)$ be the number of ways to choose r satisfactory boy-girl pairs when there are n girls and $2n-1$ boys. Let $g(n,r)$ equal the presented quantity; we need to show that $f(n,r) = g(n,r)$ for $n \geq 1$, $0 \leq r \leq n$. For $n \geq 2$ and $r \geq 1$, we consider the following cases.

(a) Girl n dances. We have $r-1$ boy-girl pairs chosen from the first $n-1$ girls and the first $2n-3$ boys, since those are the only boys the first $n-1$ girls know. There are exactly $f(n-1, r-1)$ such pairings. For each of these $f(n-1, r-1)$ pairings, girl n can dance with any of the remaining $2n - r$ boys since she knows them all. Thus, there are $(2n - r)f(n - 1, r - 1)$ possible pairings where girl n dances.

(b) Girl n doesn't dance. We have r boy-girl pairs chosen from the first $n - 1$ girls and the first $2n - 3$ boys. There are exactly $f(n - 1, r)$ such pairings.

Adding these two numbers, we have

$$f(n, r) = f(n - 1, r) + (2n - r)f(n - 1, r - 1).$$

It is easy to check that $g(n, r)$ satisfies the same recursion. Also, $f(n, 0) = g(n, 0) = 1$, $f(n, n + 1) = g(n, n + 1) = 0$, and $f(1, 1) = g(1, 1) = 1$. Thus, $f(n, r) = g(n, r)$ for all $n \geq 1$, $0 \leq r \leq n$.

2.5 Iberoamerican Olympiad

Problem 1

Ninety eight points are given on a circle. Maria and José take turns drawing a segment between two of the points which have not yet joined by a segment. The game ends when each point has been used as the endpoint of a segment at least once. The winner is the player who draws the last segment. If José goes first, who has a winning strategy?

Solution. José has the winning strategy. We label the points according to the order they are used as an end of a segment for the first time: A_1 is the first point been used and A_{98} is the last used.

We claim that the player P_1 who uses A_{96} first is the loser. If P_1 connects $A_{96}A_i$, then $i \le 97$. Hence the other player P_2 can connect $A_{97}A_{98}$ and win the game.

Since there are $\binom{95}{2} = 4465$ (an odd number of) segments connecting $A_1, \cdots, A_{95}$, José can force Maria to be the first player to use A_{96}.

Problem 2

The incircle of triangle ABC touches BC, CA, AB at D, E, F, respectively. Let AD meet the incircle again at Q. Show that the line EQ passes through the midpoint of AF if and only if $AC = BC$.

Solution. Let line EQ meet AF at M. From equal tangents and cyclic quadrilateral $EQFD$, we have

$$\angle AQM = \angle EQD = \angle EDC = \frac{\angle CAB + \angle ABC}{2}.$$

From power of a point, $MQ \cdot ME = MF^2$. Thus $AM = MF \iff MQ{\cdot}ME = MF^2 = MA^2 \iff \triangle AQM \sim \triangle EAM \iff \angle AQM = \angle EAM = \angle CAB \iff \angle CAB = \angle ABC \iff AC = BC$, as desired.

Problem 3

Find the smallest natural number n with the property that among any n distinct numbers from the set $\{1, 2, \ldots, 999\}$, one can find four distinct numbers a, b, c, d with

$$a + 2b + 3c = d.$$

Solution. Set $\{166, 167, \ldots, 999\}$ has 834 elements and does not satisfy the required property, so $n \geq 835$. Let

$$S = \{s = s_1 < s_2 < \ldots < s_{835} = t\}$$

be a 835-element subset of $\{1, 2, \ldots, 999\}$. We claim that S satisfies the required property and thus $n = 835$.

We prove our claim by contradiction. Suppose, on the contrary, that S does not satisfy the property. Therefore, $t - 3s$ can not be written in the form of $s_i + 2s_j$ for some $s_i \neq s_j \in S - \{s, t\}$. Noticing that $t \geq s + 834$ and

$$999 \geq t > D = t - 3s \geq 834 \times 3 - 2t \geq 2502 - 2 \times 999 = 504,$$

we have at least 167 ordered pairs of disjoint numbers (a, b) such that $999 > a > b \geq 1$ and $a + 2b = D$:

$$\{(D-2, 1), (D-4, 2), \ldots, (D-2i, i), \ldots, (D-334, 167)\}.$$

Since $t - 3s$ is not of the form $s_i + 2s_j$, there are at least 166 numbers not in S – one each from the 167 pairs except one possible pair that include the number s. Thus S has at most $999 - 166 = 833$ elements, a contradiction. So our assumption is false and our claim is true.

Problem 4

Around a table are seated representatives of n countries ($n \geq 2$), such that if two representatives are from the same country, their neighbors on the right are from two different countries. Determine, for each n, the maximum number of representatives.

Solution. The answer is n^2.

Let $X_1, \ldots, X_n$ denote all the countries; let x_i denote a representative from X_i. If there are $n^2 + 1$ representatives, then by Pigeonhole Principle, at least one country, say X_1 has $n + 1$ representatives and each has a right hand neighbor. Since there are only n countries, again by Pigeonhole Principle, two of those neighbors must come from the same country, a contradiction.

Now we establish the bound inductively. For $n = 2$,

$$(x_1, x_1, x_2, x_2)$$

is a legitimate arrangement. For $n = k$, Suppose that

$$(a_1, a_2, \ldots, a_{k^2})$$

is a legitimate arrangement. Then neighbor pairs

$$(x_1, x_1),\ (x_2, x_2),\ \ldots,\ (x_k, x_k)$$

each appeared exactly once. For $n = k+1$, we build a legitimate arrangement based on $(a_1, a_2, \ldots, a_{k^2})$ by replacing(x_1, x_1) by $(x_1, x_{k+1}, x_{k+1}, x_1, x_1)$; replace (x_i, x_i) by (x_i, x_{k+1}, x_i, x_i) for each $i = 2, \cdots, k$. Therefore our induction is complete and the maximum number of representatives is n^2.

Problem 5

Find the largest integer n for which there exist distinct points $P_1, P_2, \ldots, P_n$ in the plane and real numbers $r_1, r_2, \ldots, r_n$ such that the distance between P_i and P_j is $r_i + r_j$.

Solution. The answer is 4. We notice that circles centered at P_i with radius r_i must be pairwisely externally tangent to each other and this can only happen to at most 4 circles.

Problem 6

Let λ be the positive root of the equation $t^2 - 1998t - 1 = 0$. Define the sequence $x_0, x_1, \ldots$ by setting

$$x_0 = 1, \quad x_{n+1} = \lfloor \lambda x_n \rfloor \qquad (n \geq 0).$$

Find the remainder when x_{1998} is divided by 1998.

Solution. We have

$$\begin{aligned} 1998 < \lambda &= \frac{1998 + \sqrt{1998^2 + 4}}{2} \\ &= 999 + \sqrt{999^2 + 1} \\ &< 1999, \end{aligned}$$

$x_1 = 1998$, $x_2 = 1998^2$. Since $\lambda^2 - 1998\lambda - 1 = 0$,

$$\lambda = 1998 + \frac{1}{\lambda} \quad \text{and} \quad x\lambda = 1998x + \frac{x}{\lambda}$$

for all real number x. Since $x_n = \lfloor x_{n-1}\lambda \rfloor$ and x_{n-1} is an integer and λ is irrational, we have

$$x_n < x_{n-1}\lambda < x_n + 1 \quad \text{or} \quad \frac{x_n}{\lambda} < x_{n-1} < \frac{x_n + 1}{\lambda}.$$

Since $\lambda > 1998$, $\lfloor x_n/\lambda \rfloor = x_{n-1} - 1$. Therefore,

$$\begin{aligned} x_{n+1} &= \lfloor x_n \lambda \rfloor \\ &= \left\lfloor 1998x_n + \frac{x_n}{\lambda} \right\rfloor \\ &= 1998x_n + x_{n-1} - 1, \end{aligned}$$

i.e., $x_{n+1} \equiv x_{n-1} - 1 \pmod{1998}$. Therefore by induction $x_{1998} \equiv x_0 - 999 \equiv 1000 \pmod{1998}$.

2.6 Nordic Mathematics Contest

Problem 1

Find all functions from the rational numbers to the rational numbers satisfying $f(x+y)+f(x-y)=2f(x)+2f(y)$ for all rational x and y.

Solution. The only such functions are $f(x)=kx^2$ for rational k. Any such function works, since

$$\begin{aligned} f(x+y)+f(x-y) &= k(x+y)^2+k(x-y)^2 \\ &= kx^2+2kxy+ky^2+kx^2-2kxy+ky^2 \\ &= 2kx^2+2ky^2 \\ &= 2f(x)+2f(y). \end{aligned}$$

Now suppose f is any function satisfying

$$f(x+y)+f(x-y)=2f(x)+2f(y).$$

Then letting $x=y=0$ gives $2f(0)=4f(0)$, so $f(0)=0$. We will prove by induction that $f(nz)=n^2f(z)$ for any positive integer n and rational z. The claim holds for $n=0$ and $n=1$; let $n\geq 2$ and suppose the claim holds for $n-1$ and $n-2$. Then letting $x=(n-1)z$, $y=z$ in the given equation we obtain

$$\begin{aligned} f(nz)+f((n-2)z) &= f((n-1)z+z)+f((n-1)z-z) \\ &= 2f((n-1)z)+2f(z) \end{aligned}$$

so

$$\begin{aligned} f(nz) &= 2f((n-1)z)+2f(z)-f((n-2)z) \\ &= 2(n-1)^2f(z)+2f(z)-(n-2)^2f(z) \\ &= (2n^2-4n+2+2-n^2+4n-4)f(z) \\ &= n^2f(z) \end{aligned}$$

and the claim holds by induction. Letting $x=0$ in the given equation gives

$$f(y)+f(-y)=2f(0)+2f(y)=2f(y),$$

so $f(-y) = f(y)$ for all rational y; thus $f(nz) = n^2 f(z)$ for all integers n. Now let $k = f(1)$; then for any rational number $x = p/q$,

$$q^2 f(x) = f(qx) = f(p) = p^2 f(1) = kp^2$$

so

$$f(x) = kp^2/q^2 = kx^2.$$

Thus the functions $f(x) = kx^2$ are the only solutions.

Problem 2

Let C_1 and C_2 be two circles which intersect at A and B. Let M_1 be the center of C_1 and M_2 the center of C_2. Let P be a point on the segment AB such that $|AP| \neq |BP|$. Let the line through P perpendicular to M_1P meet C_1 at C and D, and let the line through P perpendicular to M_2P meet C_2 at E and F. Prove that C, D, E, F are the vertices of a rectangle.

Solution. Since CD is perpendicular to M_1P, P is the midpoint of CD. Similarly, P is the midpoint of EF, so segments CD and EF bisect each other and $CEDF$ is a parallelgram. Since P lies on AB, the radical axis of C_1 and C_2, P has equal powers with respect to these circles, so $PC \cdot PD = PE \cdot PF$ and $CD = EF$; thus $CEDF$ is in fact a rectangle.

Problem 3

(a) For which positive integers n does there exist a sequence $x_1, \ldots, x_n$ containing each of the integers $1, 2, \ldots, n$ exactly once, and such that k divides

$$x_1 + x_2 + \cdots + x_k$$

for $k = 1, 2, \ldots, n$?

(b) Does there exist an infinite sequence $x_1, x_2, \ldots$ containing every positive integer exactly once, and such that for any positive integer k, k divides

$$x_1 + x_2 + \cdots + x_k?$$

Solution. **(a)** The only such n are 1 and 3, for which we have the sequences 1 and 1, 3, 2 respectively. Suppose n is a positive integer for which there exists a permutation $x_1, \ldots, x_n$ of $1, 2, \ldots, n$ such that k

divides $x_1 + x_2 + \cdots + x_k$ for each k. Then n divides

$$x_1 + x_2 + \cdots + x_n = 1 + 2 + \cdots + n = n(n+1)/2,$$

so $(n+1)/2$ must be an integer; thus n must be odd. Now suppose $n \geq 5$. Let $m = (n+1)/2$. From

$$(n-1)|x_1 + x_2 + \cdots + x_{n-1} = nm - x_n.$$

we have

$$x_n \equiv nm \equiv m \pmod{n-1}$$

which implies that $x_n = m$ since otherwise

$$x_n \geq m + (n-1) > n \text{ or } x_n \leq m - (n-1) < 1$$

as $n \geq 3$. Similarly,

$$(n-2)|x_1 + x_2 + \cdots + x_{n-2} = (n-1)m - x_{n-1},$$

which implies that

$$x_{n-1} \equiv (n-1)m \equiv m \pmod{n-2}.$$

We obtain $x_{n-1} = m$ as

$$m + (n-2) > n \text{ and } m - (n-2) < 1$$

for $n \geq 5$. But then $x_{n-1} = m = x_n$, so the sequence $x_1, x_2, \ldots, x_n$ cannot contain each of the integers 1, 2, ..., n exactly once. Thus the only possible values of n are 1 and 3.

(b) We will construct such a sequence as follows. Given a partial sequence $x_1, x_2, \ldots, x_N$ of distinct positive integers such that k divides $x_1 + x_2 + \cdots + x_k$ for $k \leq N$, and a positive integer n, we show that this sequence can be extended to another such sequence containing n. If n is already contained in the sequence, there is nothing to show. Otherwise, let $s = x_1 + x_2 + \cdots + x_N$; since $N+1$ and $N+2$ are relatively prime, by the Chinese Remainder Theorem we can find an integer m_0 such that $m_0 \equiv -s \pmod{N+1}$ and $m_0 \equiv -s - n \pmod{N+2}$. Specifically, we can take

$$m_0 = n(N+1) - s.$$

Let $m > x_1, x_2, \ldots, x_N, n$ be an integer such that

$$m \equiv m_0 \pmod{(N+1)(N+2)};$$

take $x_{N+1} = m$ and $x_{N+2} = n$. Then $x_1, x_2, \ldots, x_{N+2}$ are all distinct,

$$(N+1)|x_1 + x_2 + \cdots + x_{N+1} = s + x_{N+1},$$

and

$$(N+2)|x_1 + x_2 + \cdots + x_{N+2} = s + x_{N+1} + n,$$

so $x_1, x_2, \ldots, x_{N+2}$ is a sequence of the type we want containing n.

Now starting with the sequence $x_1 = 1$, we succesively add $2, 3, \ldots$ in this way. This gives us a sequence $x_1, x_2, \ldots$ with the desired properties.

Problem 4

Let n be a positive integer. Prove that the number of $k \in \{0, 1, \ldots, n\}$ for which $\binom{n}{k}$ is odd is a power of 2.

Solution. Let the base 2 expansion of n be $2^0 n_0 + 2^1 n_1 + \cdots + 2^a n_a$, where $n_i \in \{0, 1\}$ for each i. Then for any $k = 2^0 k_0 + 2^1 k_1 + \cdots + 2^a k_a$, we have

$$\binom{n}{k} \equiv \binom{n_0}{k_0}\binom{n_1}{k_1}\cdots\binom{n_a}{k_a} \pmod 2$$

by Lucas' theorem. Thus $\binom{n}{k}$ is odd if and only if $k_i \le n_i$ for each i. Let m be the number of n_i's equal to 1; then the values of $k \in \{0, 1, \ldots, 2^{a+1} - 1\}$ for which $\binom{n}{k}$ is odd are obtained by setting $k_i = 0$ or 1 for each of the m values of i such that $n_i = 1$, and $k_i = 0$ for the other values of i. Thus there are 2^m values of k in $\{0, 1, \ldots, 2^{a+1} - 1\}$ for which $\binom{n}{k}$ is odd. Finally, note that for $k > n$, $\binom{n}{k} = 0$ is never odd, so the number of $k \in \{0, 1, \ldots, n\}$ for which $\binom{n}{k}$ is odd is 2^m, a power of 2.

2.7 St. Petersburg City Mathematical Olympiad (Russia)

Problem 1

In how many zeroes can the number $1^n + 2^n + 3^n + 4^n$ end for $n \in \mathbb{N}$?

Solution. There can be no zeroes (i.e., $n = 4$), one zero ($n = 1$) or two zeroes ($n = 2$). In fact, for $n \geq 3$, 2^n and 4^n are divisible by 8, while $1^n + 3^n$ is congruent to 2 or 4 mod 8. Thus the sum cannot end in 3 or more zeroes.

Problem 2

The diagonals of parallelogram $ABCD$ meet at O. The circumcircle of triangle ABO meets AD at E, and the circumcircle of DOE meets BE at F. Show that $\angle BCA = \angle FCD$.

Solution. We use directed angles. From cyclic quadrilaterals,

$$\angle EFD = \angle EOD = \angle DAB = \angle BCD$$

so $BFDC$ is also cyclic. Thus

$$\angle BCF = \angle BDF = \angle OEF = \angle OAB = \angle ACD,$$

whence $\angle BCA = \angle FCD$.

Problem 3

In a 10×10 table are written the numbers from 1 to 100. From each row we select the third largest number. Show that the sum of these numbers is not less than the sum of the numbers in some row.

Solution. Let $a_0 > \cdots > a_9$ be the numbers selected. Then at most 20 numbers exceed a_0 (the largest and second-largest in each row), so $a_0 \geq 80$. Similarly, $a_1 \geq 72$ (this time, the largest and second-largest in each row, and the elements of the row containing a_0 may exceed a_1). Hence

$$\begin{aligned} a_0 + \cdots + a_9 &\geq 80 + 72 + (a_9 + 7) + (a_9 + 6) + \cdots + a_9 \\ &= 8a_9 + 180. \end{aligned}$$

Meanwhile, the row containing a_9 has sum at most

$$100 + 99 + a_9 + \cdots + (a_9 - 7) = 8a_9 + 171,$$

which is less than the sum of the a_i.

Problem 4

Show that the projections of the intersection of the diagonals of a cyclic quadrilateral onto two opposite sides are symmetric across the line joining the midpoints of the other two sides.

Solution. Let $ABCD$ be the quadrilateral, O the intersection of its diagonals, K and L the projections of O on BC and DA, and M and N the midpoints of AB and CD. Also let P and Q be the midpoints of AO and BO. Then $MP = BO/2$ by similarity, while $KQ = BO/2$ since KQ is a median in the right triangle BKO. Thus $MP = KQ$; similarly, $MQ = LP$. Moreover, $\angle APL = \angle KQB$ from the cyclic quadrilateral (since triangles ALO and BKO are similar), and $\angle APM = \angle AOB = \angle MQB$ by parallel lines. Hence $\angle KQM = \angle LPM$ and triangles MQK and LPM are congruent, whence $MK = ML$. Likewise $NK = NL$, so K and L are symmetric across MN.

Problem 5

The set M consists of n points in the plane, no three lying on a line. For each triangle with vertices in M, count the number of points of M lying in its interior. Prove that the arithmetic mean of these numbers does not exceed $n/4$.

Solution. It suffices to show that if P_1, P_2, P_3, P_4 are four randomly chosen points of the set, then the probability that P_4 lies in $P_1P_2P_3$ is at most $1/4$. In fact, at most one of these four points lies inside the triangle formed by the other three, from which the result follows.

Problem 6

Two piles of matches lie on a large table, one containing 2^{100} matches, the other containing 3^{100} matches. Two players take turns removing matches from the piles. On a turn, a player may take k matches from one pile and m from the other, as long as $|k^2 - m^2| \leq 1000$. The player taking the last match loses. Which player wins with optimal play?

Solution. We call a position (a, b) with a matches in one pile and b in the other *winning* if the player who starts his turn on that position can win; otherwise, call that position *losing*. We claim that the first player wins; suppose by way of contradiction that $(2^{100}, 3^{100})$ is a losing position. Then no matter how the first player first moves, he leaves a winning position; specifically, $(2^{100} - 500i, 3^{100} - 500i)$ is a winning position for any $i = 1, \ldots, 2002$. From each of these positions, the second player can remove some amount of matches (k_i, m_i) from the piles to leave another losing position (c_i, d_i).

We claim that these 2002 subsequent losing positions are all distinct; suppose on the contrary that $(c_i, d_i) = (c_j, d_j)$ for some $i < j$, so that $(k_j, m_j) = (k_i + 500(j - i), m_i + 500(j - i))$. Note that $|k_i - m_i| \geq 1$, or else the first player could have left (c_i, d_i) for the second player. Then

$$\begin{aligned} k_j^2 - m_j^2| &= |k_i - m_i|(k_j + m_j) \\ &\geq |k_i - m_i|(k_i + m_i + 1000) \\ &\geq |k_i^2 - m_i^2| + 1000 > 1000, \end{aligned}$$

a contradiction.

Now for each of the 2002 positions, $-1000 \leq k_i - m_i \leq 1000$, so by the Pigeonhole Principle two of the $k_i - m_i$ are equal. Then two of the $c_i - d_i$ are equal and one losing position admits a move to a different losing position, which is impossible.

Problem 7

On a train are riding 175 passengers and 2 conductors. Each passenger buys a ticket only after the third time she is asked to do so. The conductors take turns asking a passenger who does not already have a ticket to buy one, doing so until all passengers have bought tickets. How many tickets can the conductor who goes first be sure to sell?

Solution. The first conductor can sell all of the tickets using the following strategy: if some customer has been asked twice, ask that customer, otherwise ask a customer who has not yet been asked. Note that on the first conductor's turn, the number of customers asked an odd number of times is even, so the first conductor always has a move. Moreover, since the first conductor never asks a customer for the second time, the second conductor will never have the opportunity to sell a ticket.

Problem 8

On each of 10 sheets of paper are written several powers of 2. The sum of the numbers on each sheet is the same. Show that some number appears at least 6 times among the 10 sheets.

Solution. Let N be the common sum, and n the largest integer such that $2^n \leq N$. If each power of 2 occurs at most 5 times, the sum of all of the powers of 2 written is at most

$$5(1 + 2 + \cdots + 2^n) = 5(2^{n+1} - 1) < 10N,$$

contradiction. Thus some power occurs at least 6 times.

Problem 9

A country contains 1998 cities, any two joined by a direct flight. The ticket prices on each of these flights are different. Is it possible that any two trips visiting each city once and returning to the city of origin have different total prices?

Solution. Yes. Choose the prices to be distinct powers of 2; then every subset of flights has a different total price.

Problem 10

Show that for any natural number n, between n^2 and $(n+1)^2$ one can find three distinct natural numbers a, b, c such that $a^2 + b^2$ is divisible by c.

Solution. (We must assume $n > 1$.) Take

$$a = n^2 + 2, \ b = n^2 + n + 1, \ c = n^2 + 1;$$

then $a^2 + b^2 = (2n^2 + 2n + 5)c$.

Problem 11

On a circle are marked 999 points. How many ways are there to assign to each point one of the letters A, B, or C, so that on the arc between any two points marked with the same letter, there are an even number of letters differing from these two?

Solution. There are $2^{999}+1$ such arrangements. First, any letter that occurs must occur an odd number of times, so, aside from the 3 arrangements consisting entirely of one letter, any such arrangement contains each letter at least once.

Between each pair of letters, write down the triple (a, b, c), where a denotes the number of letters one meets, counting to the left, before the next A, and similarly for b and c. Let us show that exactly two of these must be even. Since one number is 0, not all three can be odd. If just one, say a, is even, then the next letter to the right is A, and the triple beyond it has all three even. Conversely, if all three are even, the next triple to the right has exactly one even. Thus if either all even or one even occurs, these two types alternate, which is impossible because the number of triples is odd. Thus all triples must have two even terms.

Now we may write X, Y, Z in place of a triple with a odd, b odd, c odd, respectively. This gives a bijection between arrangements of the desired type and XYZ-arrangements with no two consecutive letters equal. To count these, we solve a slightly more general problem. Let a_n (resp. b_n) be the number of XYZ-sequences of length n with no two consecutive letters equal, and with the letters on the ends equal (resp. not equal). Then

$$a_{n+1}=b_n,\ b_{n+1}=2a_n+b_n(n\geq 2)$$

and by induction, $a_{2n+1}=2^{2n}+2, b_{2n+1}=2^{2n+1}-2$. The desired count is $b_{999}=2^{999}-2$, as desired.

Problem 12

A circle passing through vertices A and C of triangle ABC meets side AB at its midpoint D, and meets side BC at E. The circle through E and tangent to AC at C meets DE at F. Let K be the intersection of AC and DE. Show that the lines CF, AE, BK are concurrent.

Solution. Note that from cyclic quadrilaterals,

$$\angle EFC=\angle ECA=\angle EDB.$$

Thus FC and AB are parallel. Let G be the intersection of CF and BK, and G_1 the intersection of AE and CF. The median DK in triangle ABK bisects GC, so $FC=FG$. On the other hand, A, D, B map under a suitable homothety through E to G_1, F, C, respectively. Thus $FC=FG_1$, and $G=G_1$. Hence BK, AE, CF concur at G, as desired.

Problem 13

Can one choose 64 unit cubes from an $8 \times 8 \times 8$ cube (consisting of 8^3 unit cubes) so that any $1 \times 8 \times 8$ layer of the cube parallel to a face contains 8 of the chosen cubes, and so that among any 8 chosen cubes, two must lie in a common layer?

Solution. This is possible. Let the centers of the cubes lie at (i, j, k) with $i, j, k \in \{0, \dots, 7\}$, and choose those cubes with $i + j + k$ divisible by 8. Clearly any layer contains 8 of the cubes. On the other hand, suppose (i_m, j_m, k_m) for $m = 1, \dots, 8$ are eight chosen cubes, no two in a common layer (i.e., no two with a common coordinate). Then the sum of all of the coordinates of all eight cubes is divisible by 8; on the other hand, this sum is $3(0 + \cdots + 7) = 84$. Contradiction.

Problem 14

Find all polynomials $P(x, y)$ in two variables such that for any x and y, $P(x + y, y - x) = P(x, y)$.

Solution. Clearly any constant P works. Note that

$$P(x, y) = P(x + y, y - x) = P(2y, -2x)$$

and repeating,

$$P(x, y) = P(2y, -2x) = P(-4x, -4y) = \cdots = P(16x, 16y).$$

If $P(x, y) = \sum_{i,j \geq 0} a_{ij} x^i y^j$, then $P(x, y) = P(16x, 16y)$ forces $a_{ij} = 0$ for $i + j > 0$. Thus P must be constant.

Problem 15

A $2n$-gon $A_1 A_2 \cdots A_{2n}$ is inscribed in a circle with center O and radius 1. Show that

$$|\overrightarrow{A_1 A_2} + \cdots + \overrightarrow{A_{2n-1} A_{2n}}| \leq 2 \sin \frac{1}{2}(\angle A_1 O A_2 + \cdots + \angle A_{2n-1} O A_{2n}).$$

Solution. It suffices to prove the result when the angle on the right side is $\leq \pi/2$, since otherwise we just shift the indices. Now we proceed by induction. $n = 1$ is a trivial equality. For $n \geq 2$ we will find points A'_{2n-3}, A'_{2n-2} lying between A_{2n-3} and A_{2n} such that

$$\overrightarrow{A'_{2n-3} A'_{2n-2}} = \overrightarrow{A_{2n-3} A_{2n-2}} + \overrightarrow{A_{2n-1} A_{2n}}$$

and

$$\angle A'_{2n-3}OA'_{2n-2} \le \angle A_{2n-3}OA_{2n-2} + \angle A_{2n-1}OA_{2n};$$

applying the induction hypothesis to $A_1A_2 \dots A'_{2n-3}A'_{2n-2}$ (and the fact that sine function is increasing on $[0, \pi/2]$) gives our result. For this claim, just draw parallelogram $A_{2n-2}A_{2n-1}A_{2n}B$ and also choose C on the circle so that $A_{2n-2}C = A_{2n-1}A_{2n}$; now imagine moving the vector $\overrightarrow{A_{2n-3}B}$ around the circle until its tail reaches A_{2n}. At one end its head is inside the circle and at the other it is outside, so somewhere it must hit the circle and this gives the position of points A'_{2n-3}, A'_{2n-2}. Also the law of cosines in triangles $A_{2n-3}A_{2n-2}B$ and $A_{2n-3}A_{2n-2}C$ quickly gives $A_{2n-3}B \le A_{2n-3}C$ from which the desired angle inequality follows.

Problem 16

Let $d(n)$ denote the number of divisors of the natural number n. Prove that the sequence $d(n^2+1)$ does not become monotonic from any given point onwards.

Solution. We first note that for n even, $d(n^2+1) \le n$. Indeed, exactly half of the divisors of n^2+1 are less than n, and all are odd, so there are at most $2(n/2)$ in all.

Now if $d(n^2+1)$ becomes monotonic for $n \ge N$, then

$$d((n+1)^2+1) \ge d(n^2+1)+2$$

for $n \ge N$ (since $d(k)$ is even for k not a perfect square). Thus

$$d(n^2+1) \ge d(N^2+1) + 2(n-N)$$

which exceeds n for n large, contradiction.

Problem 17

A regiment consists of 169 men. Each day, four of them are on duty. Is it possible that at some point, any two men have served together exactly once?

Solution. Yes, this is possible. Divide the soldiers into brigades numbered $0, \dots, 12$, and within brigade k label the soldiers $(k, 0), \dots, (k, 12)$. (All labels are modulo 13.) First within each brigade, we assign soldiers m, $m+2$, $m+3$, $m+7$ to serve together, for $m = 0, \dots, 12$, so that each soldier has served with each other soldier in his brigade exactly once.

Now for each $m = 0, \ldots, 12$, we make some assignments between soldiers from brigades m, $m+2$, $m+3$, $m+7$. (Note each pair of brigades thus occurs for exactly one choice of m.) Namely, for each $n = 0, \ldots, 12$, $k = 0, \ldots, 12$, we assign soldiers $(m, n), (m+2, n+k), (m+3, n+2k), (m+7, n+3k)$ to serve. It is simple to verify that every pair of soldiers from different brigades serves together.

By now every pair has served together at least once, after exactly $13^2 + 13^3$ days with six pairs serving each day. But $6(13^2 + 13^3) = \binom{169}{2}$, which implies every pair has served together *exactly* once, as desired.

Problem 18

Can we place one of the letters P, A, S in each square of an 11×11 array so that the first row reads $PAPASPASPSA$, the letter S appears nowhere else along the perimeter, and no three-square figure shaped like an L or its 180° rotation contains three different letters?

Solution. No, it is not possible. Suppose we had such an arrangement; then the number of ways to find an A-S pair inside a single L (by which we mean an L or its rotation) is even, since each L contains either 0 or 2 such pairs. On the other hand, we can count this quantity from the point of view of each A-S pair: such a pair appearing in opposite corners of a 2×2 square is counted twice; and a pair appearing in adjacent squares is counted once if these squares lie along an edge, and twice otherwise. Thus the parity of this count is the same as the number of A-S pairs along the perimeter. But the hypotheses guarantee that this number is 3 and that the count is odd, a contradiction.

Problem 19

Around a circle are placed 20 ones and 30 twos so that no three consecutively placed numbers are equal. Find the sum of the products of every three consecutively placed numbers.

Solution. Any three consecutively placed numbers must be two 1s and a 2, or two 2s and a 1. Thus if s is their sum and p their product, we always have $p = 2s - 6$. So the sum of all 50 products p is the same as the sum of all 50 quantities $2s - 6$. Since each number appears in three sums s, this grand sum is

$$2 \times 3(20 \times 1 + 30 \times 2) - 6(50) = 180.$$

Problem 20

In triangle ABC, point K lies on AB, N on BC, and M is the midpoint of AC. It is given that $\angle BKM = \angle BNM$. Show that the perpendiculars to BC, CA, AB through N, M, K, respectively, are concurrent.

Solution. Let S be the intersection of the perpendiculars through M and K, and T the intersection of the perpendiculars through M and N. The quadrilateral $AMSK$ is cyclic, so $\angle SAM = \angle SKM = \angle BKM - \pi/2$ (or $\pi/2 - \angle BKM$ if $\angle BKM$ is acute). Analogously, $\angle TCM = \angle BNM - \pi/2$ (or $\pi/2 - \angle BNM$ if $\angle BNM$ is acute). Thus $\angle SAM = \angle TCM$, and since M is the midpoint of AC, T and S must be the same point. Thus the perpendiculars are concurrent.

Problem 21

The vertices of a connected graph are colored in 4 colors so that any edge joins vertices of different colors, and any vertex is joined to the same number of vertices of the other three colors (so if the colors were blue, red, green, and yellow, then a blue vertex joined to 7 red vertices must also be joined to 7 green vertices and 7 yellow vertices). Show that the graph remains connected if any two edges sharing a vertex are removed.

Solution. Let v be a vertex, and suppose removing two edges from v disconnects the graph. Let A_1, A_2, A_3, A_4 be the sets of vertices in the connected component of v in each color, with $v \in A_1$. Note that the numbers of edges between A_2 and any of the other groups A_1, A_3, A_4 are the same. Likewise, there are the same number of edges between A_3 and any of the other groups, and there are the same number between A_4 and any of the other groups. Thus there are the same number of edges between any two of the groups. In particular, there are the same number of edges between A_1 and any of the other three groups. However, this is impossible, because v shared two of its edges with vertices in the other connected component(s), and so the numbers of edges it shares with A_2, A_3, A_4 are different (whereas the other vertices in A_1 *do* share the same number of edges with each other group).

Problem 22

Show that any number greater than $n^4/16$ ($n \in \mathbb{N}$) can be written in at most one way as the product of two of its divisors having difference not exceeding n.

First Solution. Suppose, on the contrary, that there exist $a > c \geq d > b$ with $a-b \leq n$ and $ab = cd > n^4/16$. Put $p = a+b$, $q = a-b$, $r = c+d$, $s = c-d$. Now

$$p^2 - q^2 = 4ab = 4cd = r^2 - s^2 > n^4/4.$$

Thus $p^2 - r^2 = q^2 - s^2 \leq q^2 \leq n^2$. But $r^2 > n^4/4$ (so $r > n^2/2$) and $p > r$, so

$$p^2 - r^2 > (n^2/2+1)^2 - (n^2/2)^2 \geq n^2 + 1,$$

a contradiction.

Second Solution. Again, suppose $ab = cd > n^4/16$, with $a > c, d$ and $n \geq a - b$. If we let $p = \gcd(a, c)$, we can find positive integers p, q, r, s such that $a = pq$, $b = rs$, $c = pr$, $d = qs$. Then $a > c \Longrightarrow q > r$ and $a > d \Longrightarrow p > s$, so that

$$\begin{aligned} n &\geq pq - rs \geq (s+1)(r+1) - rs \\ &= r + s + 1 \\ &\geq 2\sqrt{b} + 1. \end{aligned}$$

Thus $b \leq (\frac{n-1}{2})^2 < n^2/4$, and $a \leq b+n \leq n^2/4$. Therefore $ab < n^4/16$, a contradiction.

Note. Compare this problem with India 1998/8.

Problem 23

A convex $2n$-gon has its vertices at lattice points. Prove that its area is not less than $n^3/100$.

Solution. Assume without loss of generality that the polygon is centrally symmetric. Otherwise, draw a main diagonal (connecting opposite vertices) to split the polygon into two smaller polygons; one of them has at most half the original area, and by joining it with its $180°$ rotation, we form a centrally symmetric, convex $2n$-gon whose area does not exceed the original polygon.

Let $\vec{w}_1, \ldots, \vec{w}_n$ be the vectors along n consecutive sides A_1A_2, A_2A_3, ..., A_nA_{n+1} of the polygon in that order; the area K these sides enclose is exactly half the polygon's area, and

$$K = [A_1A_2A_3] + [A_1A_3A_4] + \cdots + [A_1A_nA_{n+1}].$$

Since two of the vectors along the sides of triangle $A_1A_jA_{j+1}$ are $\vec{w}_j$ and $\vec{w}_1+\vec{w}_2+\cdots+\vec{w}_{j-1}$, we have the area of the polygon is

$$\begin{aligned} 2K &= 2\sum_j \frac{1}{2}|\vec{w}_j \times (\vec{w}_1+\vec{w}_2+\cdots+\vec{w}_{j-1})| \\ &= \sum_j |\vec{w}_j \times \vec{w}_0| + |\vec{w}_j \times \vec{w}_1| + \cdots + |\vec{w}_j \times \vec{w}_{j-1}| \\ &= \sum_{i<j} |\vec{w}_j \times \vec{w}_i|. \end{aligned}$$

Now, as with the $\vec{w}_i$, let $\vec{v}_1,\ldots,\vec{v}_n,-\vec{v}_1,\ldots,-\vec{v}_n$ be the vectors along the sides of the polygon, except labeled so that $|\vec{v}_1| \ge \cdots \ge |\vec{v}_n|$. Then the area S of the polygon equals

$$\sum_{i<j} |\vec{v}_i \times \vec{v}_j|.$$

Write $\vec{v}_i = a_i\vec{i}+b_i\vec{j}$ and let $k_i = \gcd(a_i,b_i)$, with $a_i = k_ix_i$ and $b_i = k_iy_i$. For a fixed $m>0$ consider the planar vectors $\vec{r}$ from the origin to a lattice point, such that $|\vec{v}_i \times \vec{r}| = m$ and $|\vec{r}| \le |\vec{v}_i|$. The set of endpoints of the vectors $\vec{r}$ are on a pair of lines parallel ℓ_1,ℓ_2 to $\vec{v}_i$, and they are not outside the circle centered at the origin with radius $|\vec{v}_i|$.

The diameter of this circle that includes $\vec{v}_i$ has exactly $2k_i+1$ lattice points: (nx_i, ny_i) for $-k \le n \le k$. Then ℓ_1,ℓ_2 each contain at most $2k_i$ lattice points inside the circle $|\vec{r}| \le |\vec{v}_i|$.

In short, for fixed i, no value of $g_i(j) = |\vec{v}_i \times \vec{v}_j|$ occurs more than $4k_i$ times for $|\vec{v}_j| \le |\vec{v}_i|$. Furthermore, since k_i divides all the components of $\vec{v}_i$, it must divide each value of $g_i(j)$. Then if we write $n-i = 4k_is+t$, with $0 \le t < 4k_i$, we conclude that

$$\begin{aligned} \sum_{j=i+1}^{n} g_i(j) &\ge 4k_i(k_i+\cdots+sk_i) + t\left((s+1)k_i\right) \\ &= k_i(2k_is+t)(s+1). \end{aligned}$$

Now

$$2k_is+t \ge 2k_is+\frac{t}{2} = \frac{n-i}{2} \quad \text{and} \quad s+1 > \frac{n-i}{4k_i}.$$

Thus

$$\sum_{j=i+1}^{n} g_i(j) > \frac{(n-i)^2}{8}.$$

Therefore

$$S \geq \sum_{i=1}^{n} \frac{(n-i)^2}{8} = \frac{(n-1)(n)(2n-1)}{48},$$

which is greater than $\frac{n^3}{100}$ for $n \geq 2$.

Problem 24

In the plane are given several vectors, the sum of whose lengths is 1. Show that they may be divided into three groups (possibly empty) so that the sum of the length of the sum of the vectors in each group is at least $3\sqrt{3}/2\pi$.

Solution. Let the vectors be $\vec{v}_1, \ldots, \vec{v}_n$, where $\vec{v}_i$ has length l_i and makes directed angle θ_i with the positive x-axis. For some angle ψ, draw rays out of the origin at angles ψ, $\psi + 2\pi/3$, $\psi + 4\pi/3$. For each vector, determine with which ray it makes the smallest (undirected) angle, and group the vectors accordingly. Now the lengths of the sum of the vectors in each group is not less than the length of its projection onto the corresponding ray. Thus, for instance, if $-\pi/3 \leq (\theta_i - \psi) \leq \pi/3$ (modulo 2π), then $\vec{v}_i$ contributes at least $l_i \cos(\theta_i - \psi)$ to the sum of the lengths of the sums. Hence, letting $\phi = \theta_i - \psi$, the average value of this length as ψ varies is at least

$$\begin{aligned}\frac{1}{2\pi} \cdot 3 \int_{-\pi/3}^{\pi/3} l_i \cos\phi \, d\phi &= \frac{3}{2\pi} l_i [\sin(\pi/3) - \sin(-\pi/3)] \\ &= \frac{3\sqrt{3}}{2\pi} l_i.\end{aligned}$$

Therefore, the average sum of all n lengths is at least

$$\frac{3\sqrt{3}}{2\pi} \sum_{i=1}^{n} l_i \geq \frac{3\sqrt{3}}{2\pi}.$$

Moreover, the integrals are not constant, so some choice of ψ gives a sum of lengths of sums strictly greater than $3\sqrt{3}/2\pi$.

Problem 25

Does there exist a nonconstant polynomial P with integer coefficients and a natural number $k > 1$ such that the numbers $P(k^n)$ are pairwise relatively prime?

Solution. Such a pair does not exist. Since P is not constant and $k > 1$, there exists $s > 0$ such that $|P(k^s)| > 1$. Let q be a prime divisor of $P(k^s)$. Then $P(k^{qs})$ is also divisible by q.

Problem 26

The point I is the incenter of triangle ABC. A circle centered at I meets BC at A_1 and A_2, CA at B_1 and B_2, and AB at C_1 and C_2, where the points occur around the circle in the order $A_1, A_2, B_1, B_2, C_1, C_2$. Let A_3, B_3, C_3 be the midpoints of the arcs A_1A_2, B_1B_2, C_1C_2, respectively. The lines A_2A_3 and B_1B_3 meet at C_4, B_2B_3 and C_1C_3 meet at A_4, and C_2C_3 and A_1A_3 meet at B_4. Prove that the lines A_3A_4, B_3B_4, C_3C_4 are concurrent.

Solution. Because A_3 is the midpoint of the arc A_1A_2, there is a circle centered at I tangent to A_1A_3 and A_2A_3. Moreover, since the arcs A_1A_2, B_1B_2, C_1C_2 are equal, the same circle is also tangent to $B_1B_3, B_2B_3, C_1C_3, C_2C_3$. Thus the hexagon $A_3C_4B_3A_4C_3B_4$ has an inscribed circle. By Brianchon's Theorem, the lines A_3A_4, B_3B_4, C_3C_4 are concurrent.

Problem 27

Given a natural number n, prove that there exists $\epsilon > 0$ such that for any n positive real numbers $a_1, a_2, \ldots, a_n$, there exists $t > 0$ such that

$$\epsilon < \{ta_1\}, \{ta_2\}, \ldots, \{ta_n\} < \frac{1}{2}.$$

(Note: $\{x\} = x - \lfloor x \rfloor$ denotes the fractional part of x.)

Solution. More generally, we prove by induction on n that for any real number $0 < r < 1$, there exists $0 < \epsilon < r$ such that for $a_1, \ldots, a_n$ any positive real numbers, there exists $t > 0$ with

$$\{ta_1\}, \ldots, \{ta_n\} \in (\epsilon, r).$$

The case $n = 1$ needs no further comment.

Assume without loss of generality that a_n is the largest of the a_i. By hypothesis, for any $r' > 0$ (which we will specify later) there exists $\epsilon' > 0$ such that for any $a_1, \ldots, a_{n-1} > 0$, there exists $t' > 0$ such that

$$\{t'a_1\}, \ldots, \{t'a_{n-1}\} \in (\epsilon', r').$$

Let N be an integer also to be specified later. A standard argument using the Pigeonhole Principle shows that one of $t'a_n, 2t'a_n, \ldots, Nt'a_n$ has fractional part in $(-1/N, 1/N)$. Let $st'a_n$ be one such term, and take $t = st' + c$ for $c = (r - 1/N)/a_n$. Then

$$ta_n \in (r - 2/N, r).$$

So we choose N such that $0 < r - 2/N$, thus making $\{ta_n\} \in (r - 2/N, r)$. Note that this choice of N makes $c > 0$ and $t > 0$, as well.

As for the other ta_i, for each i we have $k_i + \epsilon' < t'a_i < k_i + r'$ for some integer k_i, so $sk_i + s\epsilon' < st'a_i < sk_i + sr'$ and

$$\begin{aligned} sk_i + \epsilon' &< (st' + c)a_i \\ &< sk_i + sr' + \frac{a_i(r - 1/N)}{a_n} \\ &\leq sk_i + Nr' + r - 1/N. \end{aligned}$$

So we choose r' such that $Nr' - 1/N < 0$, thus making $\{ta_i\} \in (\epsilon', r)$.

Therefore, letting $\epsilon = \min\{r - 2/N, \epsilon'\}$, we have $0 < \epsilon < \{ta_1\}, \{ta_2\}, \ldots, \{ta_n\} < r$ for any choices of a_i. This completes the inductive step, and the claim is true for all natural numbers n.

Problem 28

In the plane are given several unit squares with parallel sides, such that among any n of them, there exist four having a common point. Prove that the squares can be divided into at most $n - 3$ groups, such that all of the squares in a group have a common point.

Solution. First observe that in any set of squares with parallel sides, if every two squares overlap, then every three squares overlap. This in turn implies that all the squares have a common intersection: Helly's Theorem states that if any three members of a collection of finite bounded convex sets have nonempty intersection, then all the members of the collection have a common intersection.

Second, observe that if a finite set of unit squares with parallel sides contains no $k + 1$ pairwise disjoint squares, then it can be divided into $2k - 1$ groups, each with nonempty intersection. The case $k = 1$ is covered by the previous paragraph, and we proceed from there by induction. Pick a square S whose left side is as far to the left as possible. The collection of squares *not* meeting S contains no k pairwise disjoint squares, so by the induction hypothesis it can be divided into $2k - 3$ satisfactory groups.

The remaining squares each contain either the top right or the bottom right corner of S, and so can be divided into 2 groups, each with nonempty intersection – completing the inductive step and the proof of the claim.

Now back to the original problem. Let Γ be the graph whose vertices are the squares, with an edge between two vertices if the squares have nonempty intersection. By the first observation, it suffices to show that the vertices of Γ can be divided into $n-3$ or fewer complete subgraphs. Let Γ_1 be the empty subgraph of Γ with the maximum number of vertices. Let Γ_2 be the empty subgraph of $\Gamma - \Gamma_1$ with the maximum number of vertices; and similarly, let Γ_3 be the empty subgraph of $\Gamma - \Gamma_1 - \Gamma_2$ with the maximum number of vertices. Say Γ_i has n_i vertices. If $n_1 + n_2 + n_3 \geq n$, then by Pigeonhole among any 4 vertices, 2 would be in the same Γ_i and could not be adjacent, a contradiction. Thus we must have

$$n_1 + n_2 + n_3 \leq n - 1.$$

Now, $\Gamma_1 \cup \Gamma_2$ is bipartite, so can be divided into n_1 complete subgraphs by the marriage lemma. (Otherwise, for some k there would exist k vertices in Γ_2 which form an empty subgraph with at least $n_1 - k + 1$ vertices of Γ_1, contrary to the choice of Γ_1.) Meanwhile, $\Gamma - \Gamma_1 - \Gamma_2$ can be divided into $2n_3 - 1$ complete subgraphs by the second observation. Thus we are done unless

$$n_1 + 2n_3 - 1 \geq n - 2.$$

Combining the two displayed inequalities gives $n_3 \geq n_2$; since clearly $n_1 \geq n_2 \geq n_3$, we deduce $n_2 = n_3$.

Again by the second observation, $\Gamma - \Gamma_1$ can be divided into $2n_2 - 1$ complete subgraphs. Let Ω be one of those subgraphs. Note that $\Gamma_1 \cup \Omega$ can be divided into n_1 complete subgraphs, by taking each vertex of Γ_1 together with those vertices of Ω adjacent to it. (If some vertex were left out, it could be added to Γ_1 to obtain a larger empty subgraph.) This gives a decomposition of Γ into $2n_2 - 2$ complete subgraphs forming $\Gamma - \Gamma_1 - \Omega$, and n_1 complete subgraphs forming $\Gamma_1 \cup \Omega$, for a total of

$$n_1 + 2n_2 - 2 = n_1 + n_2 + n_3 - 2 \leq n - 3$$

complete subgraphs, as desired.

3

1999 National Contests: Problems

3.1 Belarus

National Olympiad, Fourth Round

Problem 10.1 Determine all real numbers a such that the function

$$f(x) = \{ax + \sin x\}$$

is periodic. Here $\{y\}$ is the fractional part of y.

Problem 10.2 Prove that for any integer $n > 1$ the sum S of all divisors of n (including 1 and n) satisfies the inequalities

$$k\sqrt{n} < S < \sqrt{2kn},$$

where k is the number of divisors of n.

Problem 10.3 There is a 7×7 square board divided into 49 unit cells, and tiles of three types: 3×1 rectangles, 3-unit-square corners, and unit squares. Jerry has infinitely many rectangles and one corner, while Tom has only one square.

(a) Prove that Tom can put his square somewhere on the board (covering exactly one unit cell) in such a way that Jerry cannot tile the rest of the board with his tiles.

(b) Now Jerry is given another corner. Prove that no matter where Tom puts his square (covering exactly one unit cell), Jerry can tile the rest of the board with his tiles.

Problem 10.4 A circle is inscribed in the isosceles trapezoid $ABCD$. Let the circle meet diagonal AC at K and L (with K between A and L).

Find the value of

$$\frac{AL \cdot KC}{AK \cdot LC}.$$

Problem 10.5 Let P and Q be points on the side AB of the triangle ABC (with P between A and Q) such that

$$\angle ACP = \angle PCQ = \angle QCB,$$

and let AD be the angle bisector of $\angle BAC$. Line AD meets lines CP and CQ at M and N respectively. Given that $PN = CD$ and $3\angle BAC = 2\angle BCA$, prove that triangles CQD and QNB have the same area.

Problem 10.6 Show that the equation

$$\{x^3\} + \{y^3\} = \{z^3\}$$

has infinitely many rational non-integer solutions. Here $\{a\}$ is the fractional part of a.

Problem 10.7 Find all integers n and real numbers m such that the squares of an $n \times n$ board can be labelled $1, 2, \ldots, n^2$ with each number appearing exactly once in such a way that

$$(m-1)a_{ij} \leq (i+j)^2 - (i+j) \leq m a_{ij}$$

for all $1 \leq i, j \leq n$, where a_{ij} is the number placed in the intersection of the ith row and jth column.

Problem 11.1 Evalute the product

$$\prod_{k=0}^{2^{1999}} \left(4\sin^2 \frac{k\pi}{2^{2000}} - 3\right).$$

Problem 11.2 Let m and n be positive integers. Starting with the list 1, 2, 3, ..., we can form a new list of positive integers in two different ways.

(i) We first erase every mth number in the list (always starting with the first); then, in the list obtained, we erase every nth number. We call this *the first derived list.*

(ii) We first erase every nth number in the list; then, in the list obtained, we erase every mth number. We call this *the second derived list.*

Now, we call a pair (m, n) *good* if and only if the following statement is true: if some positive integer k appears in both derived lists, then it appears in the same position in each.

(a) Prove that $(2, n)$ is good for any positive integer n.

(b) Determine if there exists any good pair (m, n) such that $2 < m < n$.

Problem 11.3 Let $a_1, a_2, \ldots, a_{100}$ be an ordered set of numbers. At each move it is allowed to choose any two numbers a_n, a_m and change them to the numbers

$$\frac{a_n^2}{a_m} - \frac{n}{m}\left(\frac{a_m^2}{a_n} - a_m\right) \quad \text{and} \quad \frac{a_m^2}{a_n} - \frac{m}{n}\left(\frac{a_n^2}{a_m} - a_n\right)$$

respectively. Determine if it is possible, starting with the set with $a_i = 1/5$ for $i = 20, 40, 60, 80, 100$ and $a_i = 1$ otherwise, to obtain a set consisting of integers only.

Problem 11.4 A circle is inscribed in the trapezoid $ABCD$. Let K, L, M, N be the points of intersections of the circle with diagonals AC and BD respectively (K is between A and L and M is between B and N). Given that

$$AK \cdot LC = 16 \quad \text{and} \quad BM \cdot ND = \frac{9}{4},$$

find the radius of the circle.

Problem 11.5 Find the greatest real number k such that for any triple of positive real numbers a, b, c such that

$$kabc > a^3 + b^3 + c^3,$$

there exists a triangle with side legths a, b, c.

Problem 11.6 Find all integers x and y such that

$$x^6 + x^3y = y^3 + 2y^2.$$

Problem 11.7 Let O be the center of circle ω. Two equal chords AB and CD of ω intersect at L such that $AL > LB$ and $DL > LC$. Let M and N be points on AL and DL respectively such that $\angle ALC = 2\angle MON$. Prove that the chord of ω passing through M and N is equal to AB and CD.

IMO Selection Tests

Problem 1 Find all functions $h : \mathbb{Z} \to \mathbb{Z}$ such that

$$h(x+y) + h(xy) = h(x)h(y) + 1$$

for all x, $y \in \mathbb{Z}$.

Problem 2 Let a, b, $c \in \mathbb{Q}$, $ac \neq 0$. Given that the equation

$$ax^2 + bxy + cy^2 = 0$$

has a non-zero solution of the form

$$(x, y) = (a_0 + a_1\sqrt[3]{2} + a_2\sqrt[3]{4}, b_0 + b_1\sqrt[3]{2} + b_2\sqrt[3]{4})$$

with a_i, $b_i \in \mathbb{Q}$, $i = 0$, 1, 2, prove that it has also has a non-zero rational solution.

Problem 3 Suppose a and b are positive integers such that the product of all divisors of a (including 1 and a) is equal to the product of all divisors of b (including 1 and b). Does it follow that $a = b$?

Problem 4 Let a, b, c be positive real numbers such that $a^2+b^2+c^2 = 3$. Prove that

$$\frac{1}{1+ab} + \frac{1}{1+bc} + \frac{1}{1+ca} \geq \frac{3}{2}.$$

Problem 5 Suppose T_1 is similar to T_2, and the lengths of two sides and the angle between them of triangle T_1 are proportional to the lengths of two sides and the angle between them of triangle T_2 (but not necessarily the corresponding ones). Must T_1 be congruent to T_2?

Problem 6 Two real sequences $x_1, x_2, \ldots$, and $y_1, y_2, \ldots$, are defined in the following way: $x_1 = y_1 = \sqrt{3}$,

$$x_{n+1} = x_n + \sqrt{1 + x_n^2} \quad \text{and} \quad y_{n+1} = \frac{y_n}{1 + \sqrt{1 + y_n^2}}$$

for all $n \geq 1$. Prove that $2 < x_n y_n < 3$ for all $n > 1$.

Problem 7 Let O be the center of the excircle of triangle ABC opposite A. Let M be the midpoint of AC, and let P be the intersection of MO and BC. Prove that if $\angle BAC = 2\angle ACB$, then $AB = BP$.

Problem 8 Let O, O_1 be the centers of the incircle and the excircle opposite A of triangle ABC. The perpendicular bisector of OO_1 meets lines AB and AC at L and N respectively. Given that the circumcircle of triangle ABC touches line LN, prove that triangle ABC is isosceles.

Problem 9 Does there exist a bijection f of

(a) a plane

(b) three-dimensional space

such that for any distinct points A, B line AB and line $f(A)f(B)$ are perpendicular?

Problem 10 A word is a finite sequence of two symbols a and b. The number of the symbols in the word is said to be the length of the word. A word is called 6-*aperiodic* if it does not contain a subword of the form $cccccc$ for any word c. Prove that $f(n) > (3/2)^n$, where $f(n)$ is the total number of 6-aperiodic words of length n.

Problem 11 Determine all positive integers n, $n \geq 2$, such that $\binom{n-k}{k}$ is even for $k = 1, 2, \ldots, \lfloor n/2 \rfloor$.

Problem 12 A number of n players took part in a chess tournament. After the tournament was over, it turned out that among any four players there was one who scored differently against the other three (i.e., he got a victory, a draw, and a loss). Prove that the largest possible n satisfies the inequality $6 \leq n \leq 9$.

3.2 Brazil

Problem 1 Let $ABCDE$ be a regular pentagon such that the star region $ACEBD$ has area 1. Let AC and BE meet at P, and let BD and CE meet at Q. Determine $[APQD]$.

Problem 2 Given a 10×10 board, we want to remove n of the 100 squares so that no 4 of the remaining squares form the corners of a rectangle with sides parallel to the sides of the board. Determine the minimum value of n.

Problem 3 The planet Zork is spherical and has several cities. Given any city A on Zork, there exists an antipodal city A' (i.e., symmetric with respect to the center of the sphere). In Zork, there are roads joining pairs of cities. If there is a road joining cities P and Q, then there is a road joining P' and Q'. Roads don't cross each other, and any given pair of cities is connected by some sequence of roads. Each city is assigned a value, and the difference between the values of every pair of connected cities is at most 100. Prove that there exist two antipodal cities with values differing by at most 100.

Problem 4 In Tumbolia there are n soccer teams. We want to organize a championship such that each team plays exactly once with each other team. All games take place on Sundays, and a team can't play more than one game in the same day. Determine the smallest positive integer m for which it is possible to realize such a championship in m Sundays.

Problem 5 Given a triangle ABC, show how to construct, with straightedge and compass, a triangle $A'B'C'$ with minimal area such that A', B', C' lie on AB, BC, CA, respectively, $\angle B'A'C' = \angle BAC$, and $\angle A'C'B' = \angle ACB$.

3.3 Bulgaria

National Olympiad, Third Round

Problem 1 Find all triples (x, y, z) of natural numbers such that y is a prime number, y and 3 do not divide z, and $x^3 - y^3 = z^2$.

Problem 2 A convex quadrilateral $ABCD$ is inscribed in a circle whose center O is inside the quadrilateral. Let $MNPQ$ be the quadrilateral whose vertices are the projections of the intersection point of the diagonals AC and BD onto the sides of $ABCD$. Prove that $2[MNPQ] \le [ABCD]$.

Problem 3 In a competition 8 judges marked the contestants by *pass* or *fail*. It is known that for any two contestants, two judges marked both with *pass*; two judges marked the first contestant with *pass* and the second contestant with *fail*; two judges marked the first contestant with *fail* and the second contestant with *pass*; and finally, two judges marked both with *fail*. What is the largest possible number of contestants?

Problem 4 Find all pairs (x, y) of integer numbers such that

$$x^3 = y^3 + 2y^2 + 1.$$

Problem 5 Let B_1 and C_1 be points on the sides AC and AB of triangle ABC. Lines BB_1 and CC_1 intersect at point D. Prove that the quadrilateral AB_1DC_1 is circumscribed if and only if the incircles of the triangles ABD and ACD are tangent to each other.

Problem 6 Each interior point of an equilateral triangle of side 1 lies in one of six congruent circles of radius r. Prove that

$$r \ge \frac{\sqrt{3}}{10}.$$

National Olympiad, Fourth Round

Problem 1 A rectangular parallelepiped has integer dimensions. All of its faces of are painted green. The parallelepiped is partitioned into unit cubes by planes parallel to its faces. Find all possible measurements of the parallelepiped if the number of cubes without a green face is one third of the total number of cubes.

Problem 2 Let $\{a_n\}$ be a sequence of integers such that for $n \geq 1$

$$(n-1)a_{n+1} = (n+1)a_n - 2(n-1).$$

If 2000 divides a_{1999}, find the smallest $n \geq 2$ such that 2000 divides a_n.

Problem 3 The vertices of a triangle have integer coordinates and one of its sides is of length $\sqrt{n}$, where n is a square-free natural number. Prove that the ratio of the circumradius to the inradius of the triangle is an irrational number.

Problem 4 Find the number of all natural numbers n, $4 \leq n \leq 1023$, whose binary representations do not contain three consecutive equal digits.

Problem 5 The vertices A, B and C of an acute-angled triangle ABC lie on the sides B_1C_1, C_1A_1 and A_1B_1 of triangle $A_1B_1C_1$ such that $\angle ABC = \angle A_1B_1C_1$, $\angle BCA = \angle B_1C_1A_1$, and $\angle CAB = \angle C_1A_1B_1$. Prove that the orthocenters of the triangle ABC and the triangle $A_1B_1C_1$ are equidistant from the circumcenter of the triangle ABC.

Problem 6 Prove that the equation

$$x^3 + y^3 + z^3 + t^3 = 1999$$

has infinitely many integral solutions.

3.4 Canada

Problem 1 Find all real solutions to the equation $4x^2 - 40[x] + 51 = 0$, where $[x]$ denotes the greatest integer less than or equal to x.

Problem 2 Let ABC be an equilateral triangle of altitude 1. A circle, with radius 1 and center on the same side of AB as C, rolls along the segment AB. Prove that the length of the arc of the circle that is inside the triangle remains constant.

Problem 3 Determine all positive integers n such that $n = d(n)^2$, where $d(n)$ denotes the number of positive divisors of n (including 1 and n).

Problem 4 Suppose $a_1, a_2, \cdots, a_8$ are eight distinct integers from the set $S = \{1, 2, \ldots, 17\}$. Show that there exists an integer $k > 0$ such that the equation $a_i - a_j = k$ has at least three different solutions. Also, find a specific set of 7 distinct integers $\{b_1, b_2, \cdots, b_7\}$ from S such that the equation

$$b_i - b_j = k$$

does not have three distinct solutions for any $k > 0$.

Problem 5 Let x, y, z be non-negative real numbers such that

$$x + y + z = 1.$$

Prove that

$$x^2y + y^2z + z^2x \leq \frac{4}{27}.$$

and determine when equality occurs.

3.5 China

Problem 1 Let ABC be an acute triangle with $\angle C > \angle B$. Let D be a point on BC such that $\angle ADB$ is obtuse, and let H be the orthocenter of triangle ABD. Suppose that F is a point inside triangle ABC and is on the circumcircle of triangle ABD. Prove that F is the orthocenter of triangle ABC if and only if $HD \parallel CF$ and H is on the circumcircle of triangle ABC.

Problem 2 Let a be a real number. Let $\{f_n(x)\}$ be a sequence of polynomials such that $f_0(x) = 1$ and $f_{n+1}(x) = xf_n(x) + f_n(ax)$ for $n = 0, 1, 2, \ldots$.

(a) Prove that

$$f_n(x) = x^n f_n\left(\frac{1}{x}\right)$$

for $n = 0, 1, 2, \ldots$.

(b) Find an explicit expression for $f_n(x)$.

Problem 3 There are 99 space stations. Each pair of space stations is connected by a tunnel. There are 99 two-way main tunnels, and all the other tunnels are strictly one way tunnels. A group of 4 space stations is called *connected* if one can reach each station in the group from every other station in the group without using any tunnels other than the 6 tunnels which connect them. Determine the maximum number of connected groups.

Problem 4 Let m be a positive integer. Prove that there are integers a, b, k, such that both a and b are odd, $k \geq 0$ and

$$2m = a^{19} + b^{99} + k \cdot 2^{1999}.$$

Problem 5 Determine the maximum value of λ such that if $f(x) = x^3 + ax^2 + bx + c$ is a cubic polynomial with all its roots nonnegative, then

$$f(x) \geq \lambda(x - a)^3$$

for all $x \geq 0$. Find the equality condition.

Problem 6 A $4 \times 4 \times 4$ cube is composed of 64 unit cubes. The faces of 16 unit cubes are to be colored red. A coloring is called *interesting* if there is exactly 1 red unit cube in every $1 \times 1 \times 4$ rectangular box composed of 4 unit cubes. Determine the number of interesting colorings.

3.6 Czech and Slovak Republics

Problem 1 In the fraction

$$\frac{29 \div 28 \div 27 \div \cdots \div 16}{15 \div 14 \div 13 \div \cdots \div 2}$$

parentheses may be repeatedly placed anywhere in the numerator, granted they are also placed on the identical locations in the denominator.

(a) Find the least possible integral value of the resulting expression.

(b) Find all possible integral values of the resulting expression.

Problem 2 In a tetrahedron $ABCD$ we denote by E and F the midpoints of the medians from the vertices A and D, respectively. (The median from a vertex of a tetrahedron is the segment connecting the vertex and the centroid of the opposite face.) Determine the ratio of the volumes of tetrahedrons $BCEF$ and $ABCD$.

Problem 3 Show that there exists a triangle ABC for which, with the usual labelling of sides and medians, it is true that $a \neq b$ and $a + m_a = b + m_b$. Show further that there exists a number k such that for each such triangle $a + m_a = b + m_b = k(a + b)$. Finally, find all possible ratios $a : b$ of the sides of these triangles.

Problem 4 In a certain language there are only two letters, A and B. The words of this language satisfy the following axioms:

(i) There are no words of length 1, and the only words of length 2 are AB and BB.

(ii) A sequence of letters of length $n > 2$ is a word if and only if it can be created from some word of length less than n by the following construction: all letters A in the existing word are left unchanged, while each letter B is replaced by some word. (While performing this operation, the B's do not all have to be replaced by the same word.)

Show that for any n the number of words of length n equals

$$\frac{2^n + 2 \cdot (-1)^n}{3}.$$

Problem 5 In the plane an acute angle APX is given. Show how to construct a square $ABCD$ such that P lies on side BC and P lies on the bisector of angle BAQ where Q is the intersection of ray PX with CD.

Problem 6 Find all pairs of real numbers a and b such that the system of equations

$$\frac{x+y}{x^2+y^2}=a, \qquad \frac{x^3+y^3}{x^2+y^2}=b$$

has a solution in real numbers (x, y).

3.7 France

Problem 1

(a) What is the maximum volume of a cylinder that is inside a given cone and has the same axis of revolution as the cone?

(b) What is the maximum volume of a ball that is inside a given cone?

(c) Compare the two maxima you found.

Problem 2 Find all integer solutions to $(n+3)^n = \sum_{k=3}^{n+2} k^n$.

Problem 3 For which acute-angled triangle is the ratio of the shortest side to the inradius maximal?

Problem 4 There are 1999 red candies and 6661 yellow candies on a table, made indistinguishable by their wrappers. A *gourmand* applies the following algorithm until the candies are gone:

(a) If there are candies left, he takes one at random, notes its color, eats it, and goes to (b).

(b) If there are candies left, he takes one at random, notes its color, and

  (i) if it matches the last one eaten, he eats it also and returns to (b).

  (ii) if it does not match the last one eaten, he wraps it up again, puts it back, and goes to (a).

Prove that all the candies will eventually be eaten. Find the probability that the last candy eaten is red.

Problem 5 With a given triangle, form three new points by reflecting each vertex about the opposite side. Show that these three new points are collinear if and only if the the distance between the orthocenter and the circumcenter of the triangle is equal to the diameter of the circumcircle of the triangle.

3.8 Hungary

Problem 1 I have $n \geq 5$ real numbers with the following properties:

(i) They are nonzero, but at least one of them is 1999.

(ii) Any four of them can be rearranged to form a geometric progression.

What are my numbers?

Problem 2 Let ABC be a right triangle with $\angle C = 90°$. Two squares S_1 and S_2 are inscribed in triangle ABC such that S_1 and ABC share a common vertex C, and S_2 has one of its sides on AB. Suppose that $[S_1] = 441$ and $[S_2] = 440$. Calculate $AC + BC$.

Problem 3 Let O and K be the centers of the respective spheres tangent to the sides, and the edges of a right pyramid, whose base is a 2 by 2 square. Determine the volume of the pyramid if O and K are equidistant from the base.

Problem 4 For any given positive integer n, determine (as a function of n) the number of ordered pairs (x, y) of positive integers such that

$$x^2 - y^2 = 10^2 \cdot 30^{2n}.$$

Further prove that the number of such pairs is never a perfect square.

Problem 5 For $0 \leq x, y, z \leq 1$, find all solutions to the equation

$$\frac{x}{1+y+zx} + \frac{y}{1+z+xy} + \frac{z}{1+x+yz} = \frac{3}{x+y+z}.$$

Problem 6 The midpoints of the edges of a tetrahedron lie on a sphere. What is the maximum volume of the tetrahedron?

Problem 7 A positive integer is written in each square of an n^2 by n^2 chess board. The difference between the numbers in any two adjacent squares (sharing an edge) is less than or equal to n. Prove that at least $\lfloor n/2 \rfloor + 1$ squares contain the same number.

Problem 8 On some day in the 20-th century Alex noticed that adding the four digits of the year of his birth gave his actual age. On the same day, Bernath noticed also this. They are not the same age, and both are under 99. By how many years do their ages differ?

Problem 9 Let ABC be a triangle and D a point on the side AB. The incircles of the triangles ACD and CDB touch each other on CD. Prove that the incircle of ABC touches AB at D.

Problem 10 Let R be the circumradius of a right pyramid with a square base. Let r be the radius of the sphere touching the four faces. Suppose that $2R = (1 + \sqrt{2})r$. Determine the angle between adjacent faces of the pyramid.

Problem 11 Is there a polynomial $P(x)$ with integer coefficients such that $P(10) = 400$, $P(14) = 440$, and $P(18) = 520$?

Problem 12 Let a, b, c be positive numbers and $n \geq 2$ be an integer such that $a^n + b^n = c^n$. For which positive integers k is it possible to construct an obtuse triangle with sides a^k, b^k, c^k?

Problem 13 Let $n > 1$ be an arbitrary real number and k be the number of positive primes less than or equal to n. Select $k + 1$ positive integers such that none of them divides the product of all the others. Prove that there exists a number among the $k + 1$ chosen numbers which is bigger than n.

Problem 14 The polynomial $x^4 - 2x^2 + ax + b$ has four distinct real roots. Show that the absolute value of each root is smaller than $\sqrt{3}$.

Problem 15 Each side of a convex polygon has integral length and the perimeter is odd. Prove that the area of the polygon is at least $\sqrt{3}/4$.

Problem 16 Determine if there exists an infinite sequence of positive integers such that

(i) no term divides any other term.

(ii) every pair of terms has a common divisor greater than 1, but no integer greater than 1 divides all of the terms.

Problem 17 Prove that, for every positive integer n, there exists a polynomial with integer coefficients whose values at $1, 2, \ldots, n$ are different powers of 2.

Problem 18 Find all integers $N \geq 3$ for which it is possible to choose N points in the plane (no three collinear) such that each triangle formed by three vertices on the convex hull of the points contains exactly one of the points in its interior.

3.9 Iran

First Round

Problem 1 Suppose that $a_1 < a_2 < \cdots < a_n$ are real numbers. Prove that

$$a_1 a_2^4 + a_2 a_3^4 + \cdots + a_n a_1^4 \geq a_2 a_1^4 + a_3 a_2^4 + \cdots + a_1 a_n^4.$$

Problem 2 Suppose that n is a positive integer. The n-tuple $(a_1, \ldots, a_n)$ of positive integers is said to be *good* if $a_1 + \cdots + a_n = 2n$ if for every k between 1 and n, no k of the n integers add up to n. Find all n-tuples that are good.

Problem 3 Let I be the incenter of triangle ABC and let AI meet the circumcircle of ABC at D. Denote the feet of the perpendiculars from I to BD and CD by E and F, respectively. If $IE + IF = AD/2$, calculate $\angle BAC$.

Problem 4 Let ABC be a triangle with $BC > CA > AB$. Choose points D on BC and E on the extension of BA such that

$$BD = BE = AC.$$

The circumcircle of triangle BED intersects AC at P and the line BP intersects the circumcircle of triangle ABC again at Q. Prove that $AQ + QC = BP$.

Problem 5 Suppose that n is a positive integer and let

$$d_1 < d_2 < d_3 < d_4$$

be the four smallest positive integer divisors of n. Find all integers n such that

$$n = d_1^2 + d_2^2 + d_3^2 + d_4^2.$$

Problem 6 Suppose that $A = (a_1, a_2, \ldots, a_n)$ and $B = (b_1, b_2, \ldots, b_n)$ are two 0-1 sequences. The difference between A and B is defined to be the number of i's for which $a_i \neq b_i$ $(1 \leq i \leq n)$, and denoted by $d(A, B)$. Suppose that A, B, C are three 0-1 sequences and that $d(A, B) = d(A, C) = d(B, C) = d$.

(a) Prove that d is even.

(b) Prove that there exists an 0-1 sequence D such that

$$d(D,A) = d(D,B) = d(D,C) = \frac{d}{2}.$$

Second Round

Problem 1 Define the sequence $\{x_n\}_{n\geq 0}$ by $x_0 = 0$ and

$$x_n = \begin{cases} x_{n-1} + \dfrac{3^{r+1}-1}{2}, & \text{if } n = 3^r(3k+1), \\ x_{n-1} - \dfrac{3^{r+1}+1}{2}, & \text{if } n = 3^r(3k+2), \end{cases}$$

where k and r are nonnegative integers. Prove that every integer appears exactly once in this sequence.

Problem 2 Suppose that $n(r)$ denotes the number of points with integer coordinates on a circle of radius $r > 1$. Prove that

$$n(r) < 6\sqrt[3]{\pi r^2}.$$

Problem 3 Find all functions $f : \mathbb{R} \to \mathbb{R}$ satisfying

$$f(f(x) + y) = f(x^2 - y) + 4f(x)y$$

for all $x, y \in \mathbb{R}$.

Problem 4 In triangle ABC, the angle bisector of $\angle BAC$ meets BC at D. Suppose that ω is the circle which is tangent to BC at D and passes through A. Let M be the second point of intersection of ω and AC. Let P be the second point of intersection of ω and BM. Prove that P lies on a median of triangle ABD.

Problem 5 Let ABC be a triangle. If we paint the points of the plane in red and green, prove that either there exist two red points which are one unit apart or three green points forming a triangle congruent to ABC.

Third Round

Problem 1 Suppose that $S = \{1, 2, \ldots, n\}$ and that $A_1, A_2, \ldots, A_k$ are subsets of S such that for every $1 \leq i_1, i_2, i_3, i_4 \leq k$, we have

$$|A_{i_1} \cup A_{i_2} \cup A_{i_3} \cup A_{i_4}| \leq n - 2.$$

Prove that $k \leq 2^{n-2}$.

Problem 2 Let ABC be a triangle and let ω be a circle passing through A and C. Sides AB and BC meet ω again at D and E, respectively. Let γ be the incircle of the circular triangle EBD and let S be its center. Suppose that γ touches the arc DE at M. Prove that the angle bisector of $\angle AMC$ passes through the incenter of triangle ABC.

Problem 3 Suppose that $C_1, C_2, \ldots, C_n$ are circles of radius 1 in the plane such that no two of them are tangent and the subset of the plane formed by the union of these circles is connected (i.e., for any partition of $\{1, 2, \ldots, n\}$ into nonempty subsets A and B, $\bigcup_{a\in A} C_a$ and $\bigcup_{b\in B} C_b$ are not disjoint). Prove that $|S| \geq n$, where

$$S = \bigcup_{1\leq i<j\leq n} \{C_i \cap C_j\}.$$

Problem 4 Suppose that $-1 \leq x_1, x_2, \ldots, x_n \leq 1$ are real numbers such that $x_1 + x_2 + \ldots + x_n = 0$. Prove that there exists a permutation σ such that, for every $1 \leq p \leq q \leq n$,

$$|x_{\sigma(p)} + \cdots + x_{\sigma(q)}| \leq 2 - \frac{1}{n}.$$

Prove that the expression on the right hand side cannot be replaced by $2 - \frac{4}{n}$.

Problem 5 Suppose that $r_1, \ldots, r_n$ are real numbers. Prove that there exists $S \subseteq \{1, 2, \ldots, n\}$ such that

$$1 \leq |S \cap \{i, i+1, i+2\}| \leq 2,$$

for $1 \leq i \leq n-2$, and

$$\left|\sum_{i\in S} r_i\right| \geq \frac{1}{6}\sum_{i=1}^{n} |r_i|.$$

3.10 Ireland

Problem 1 Find all the real values of x which satisfy

$$\frac{x^2}{(x+1-\sqrt{x+1})^2} < \frac{x^2+3x+18}{(x+1)^2}.$$

Problem 2 Show that there is a positive number in the Fibonacci sequence which is divisible by 1000.

Problem 3 Let D, E, F be points on the sides BC, CA, AB, respectively, of triangle ABC such that $AD \perp BC$, $AF = FB$, and BE is the angle bisector of $\angle B$. Prove that AD, BE, CF are concurrent if and only if

$$a^2(a-c) = (b^2 - c^2)(a+c),$$

where $a = BC, b = CA, c = AB$.

Problem 4 A 100 by 100 square floor is to be tiled. The only available tiles are rectangular 1 by 3 tiles, fitting exactly over three squares of the floor.

(a) If a 2 by 2 square is removed from the center of the floor, prove that the remaining part of the floor can be tiled with available tiles.

(b) If, instead, a 2 by 2 square is removed from the corner, prove that the remaining part of the floor cannot be tiled with the available tiles.

Problem 5 Define a sequence u_n, $n = 0, 1, 2, \ldots$ as follows: $u_0 = 0$, $u_1 = 1$, and for each $n \geq 1$, u_{n+1} is the smallest positive integer such that $u_{n+1} > u_n$ and $\{u_0, u_1, \ldots, u_{n+1}\}$ contains no three elements which are in arithmetic progression. Find u_{100}.

Problem 6 Solve the system of equations

$$y^2 - (x+8)(x^2+2) = 0$$

$$y^2 - (8+4x)y + (16+16x-5x^2) = 0.$$

Problem 7 A function $f : \mathbb{N} \to \mathbb{N}$ satisfies

(i) $f(ab) = f(a)f(b)$ whenever the greatest common divisor of a and b is 1;

(ii) $f(p+q) = f(p) + f(q)$ for all prime numbers p and q.

Prove that $f(2) = 2$, $f(3) = 3$, and $f(1999) = 1999$.

Problem 8 Let a, b, c, d be positive real numbers whose sum is 1. Prove that

$$\frac{a^2}{a+b}+\frac{b^2}{b+c}+\frac{c^2}{c+d}+\frac{d^2}{d+a}\geq\frac{1}{2}$$

with equality if and only if $a=b=c=d=1/4$.

Problem 9 Find all positive integers m such that the fourth power of the number of positive divisors of m equals m.

Problem 10 Let $ABCDEF$ be a convex hexagon such that $AB=BC$, $CD=DE$, $EF=FA$, and

$$\angle ABC+\angle CDE+\angle EFA=360^\circ.$$

Prove that the respective perpendiculars from A, C, E to FB, BD, DF are concurrent.

3.11 Italy

Problem 1 Given a rectangular sheet with sides a and b, with $a > b$, fold it along a diagonal. Determine the area of the triangle that passes over the edge of the paper.

Problem 2 A positive integer is said to be *balanced* if the number of its decimal digits equals the number of its distinct prime factors (for instance 15 is balanced, while 49 is not). Prove that there are only finitely many balanced numbers.

Problem 3 Let $\omega, \omega_1, \omega_2$ be three circles with radii r, r_1, r_2, respectively, with $0 < r_1 < r_2 < r$. The circles ω_1 and ω_2 are internally tangent to ω at two distinct points A and B and meet in two distinct points. Prove that AB contains an intersection point of ω_1 and ω_2 if and only if $r_1 + r_2 = r$.

Problem 4 Albert and Barbara play the following game. On a table there are 1999 sticks: each player in turn must remove from the table some sticks, provided that the player removes at least one stick and at most half of the sticks remaining on the table. The player who leaves just one stick on the table loses the game. Barbara moves first. Determine for which of the players there exists a winning strategy.

Problem 5 On a lake there is a village of pile-built dwellings, set on the nodes of an $m \times n$ rectangular array. Each dwelling is an endpoint of exactly p bridges which connect the dwelling with one or more of the adjacent dwellings (here adjacent means with respect to the array, hence diagonal connection is not allowed). Determine for which values of m, n, p it is possible to place the bridges so that from any dwelling one can reach any other dwelling. (Clearly, two adjacent dwellings can be connected by more than one bridge).

Problem 6 Determine all triples (x, k, n) of positive integers such that

$$3^k - 1 = x^n.$$

Problem 7 Prove that for each prime p the equation

$$2^p + 3^p = a^n$$

has no integer solutions (a, n) with $a, n > 1$.

Problem 8 Points D and E are given on the sides AB and AC of triangle ABC such that $DE \parallel BC$ and DE is tangent to the incircle of ABC. Prove that

$$DE \leq \frac{AB + BC + CA}{8}.$$

Problem 9

(a) Find all the strictly monotonic functions $f : \mathbb{R} \to \mathbb{R}$ such that

$$f(x + f(y)) = f(x) + y, \qquad \text{for all } x, y \in \mathbb{R}.$$

(b) Prove that for every integer $n > 1$ there do not exist strictly monotonic functions $f : \mathbb{R} \to \mathbb{R}$ such that

$$f(x + f(y)) = f(x) + y^n, \qquad \text{for all } x, y \in \mathbb{R}.$$

Problem 10 Let X be a set with $|X| = n$, and let $A_1, A_2, \ldots, A_m$ be subsets of X such that

(a) $|A_i| = 3$ for $i = 1, 2, \ldots, m$.

(b) $|A_i \cap A_j| \leq 1$ for all $i \neq j$.

Prove that there exists a subset of X with at least $\lfloor \sqrt{2n} \rfloor$ elements, which does not contain A_i for $i = 1, 2, \ldots, m$.

3.12 Japan

Problem 1 You can place a stone at each of 1999×1999 squares on a grid pattern. Find the minimum number of stones you must place such that, when an arbitrary blank square is selected, the total number of stones placed in the corresponding row and column is at least 1999.

Problem 2 Let $f(x) = x^3 + 17$. Prove that for each natural number n, $n \geq 2$, there is a natural number x for which $f(x)$ is divisible by 3^n but not by 3^{n+1}.

Problem 3 From a set of $2n+1$ weights (where n is a natural number), if any one weight is excluded, then the remaining $2n$ weights can be divided into two sets of n weights that balance each other. Prove that all the weights are equal.

Problem 4 Prove that

$$f(x) = (x^2 + 1^2)(x^2 + 2^2)(x^2 + 3^2) \cdots (x^2 + n^2) + 1$$

cannot be expressed as a product of two integral-coefficient polynomials with degree greater than 0.

Problem 5 For a convex hexagon $ABCDEF$ whose side lengths are all 1, let M and m be the maximum and minimum values of the three diagonals AD, BE, and CF. Find all possible values of m and M.

3.13 Korea

Problem 1 Let R and r be the circumradius and inradius of triangle ABC respectively, and let R' and r' be the circumradius and inradius of triangle $A'B'C'$ respectively. Prove that if $\angle C = \angle C'$ and $Rr' = R'r$, then the triangles are similar.

Problem 2 Suppose $f(x)$ is a function satisfying

$$|f(m+n) - f(m)| \leq \frac{n}{m}$$

for all rational numbers n and m. Show that for all positive integers k

$$\sum_{i=1}^{k} |f(2^k) - f(2^i)| \leq \frac{k(k-1)}{2}.$$

Problem 3 Find all positive integers n such that $2^n - 1$ is a multiple of 3 and $(2^n - 1)/3$ is a divisor of $4m^2 + 1$ for some integer m.

Problem 4 Suppose that for any real x with $|x| \neq 1$, a function $f(x)$ satisfies

$$f\left(\frac{x-3}{x+1}\right) + f\left(\frac{3+x}{1-x}\right) = x.$$

Find all possible $f(x)$.

Problem 5 Consider a permutation $a_1a_2a_3a_4a_5a_6$ of 6 numbers

$$\{1, 2, 3, 4, 5, 6\}$$

which can be transformed to 123456 by transposing two numbers exactly 4 times. Find the number of such permutations.

Problem 6 Let $a_1, a_2, \cdots, a_{1999}$ be nonnegative real numbers satisfying the following two conditions:

(a) $a_1 + a_2 + \cdots + a_{1999} = 2$;

(b) $a_1a_2 + a_2a_3 + \cdots + a_{1998}a_{1999} + a_{1999}a_1 = 1$.

Let $S = a_1^2 + a_2^2 + \cdots + a_{1999}^2$. Find the maximum and minimum possible values of S.

3.14 Poland

Problem 1 Let D be a point on side BC of triangle ABC such that $AD > BC$. Point E on side AC is defined by the equation

$$\frac{AE}{EC} = \frac{BD}{AD - BC}.$$

Show that $AD > BE$.

Problem 2 Given are nonnegative integers $a_1 < a_2 < \cdots < a_{101}$ smaller than 5050. Show that one can choose four distinct integers a_k, a_l, a_m, a_n such that

$$5050|(a_k + a_l - a_m - a_n).$$

Problem 3 For a positive integer n, let $S(n)$ denote the sum of its digits. Prove that there exist distinct positive integers $\{n_i\}_{1 \le i \le 50}$ such that

$$n_1 + S(n_1) = n_2 + S(n_2) = \cdots = n_{50} + S(n_{50}).$$

Problem 4 Find all integers $n \geq 2$ for which the system of equations

$$\begin{aligned} x_1^2 + x_2^2 + 50 &= 16x_1 + 12x_2 \\ x_2^2 + x_3^2 + 50 &= 16x_2 + 12x_3 \\ &\cdots\cdots\cdots \\ x_{n-1}^2 + x_n^2 + 50 &= 16x_{n-1} + 12x_n \\ x_n^2 + x_1^2 + 50 &= 16x_n + 12x_1 \end{aligned}$$

has a solution in integers $(x_1, x_2, \ldots, x_n)$.

Problem 5 Let $a_1, a_2, \ldots, a_n, b_1, b_2, \ldots, b_n$ be integers. Prove that

$$\sum_{1 \le i < j \le n} (|a_i - a_j| + |b_i - b_j|) \le \sum_{1 \le i, j \le n} |a_i - b_j|.$$

Problem 6 In a convex hexagon $ABCDEF$, $\angle A + \angle C + \angle E = 360°$ and

$$AB \cdot CD \cdot EF = BC \cdot DE \cdot FA.$$

Prove that $AB \cdot FD \cdot EC = BF \cdot DE \cdot CA$.

3.15 Romania

National Olympiad

Problem 7.1 Determine the side lengths of a right triangle if they are integers and the product of the leg lengths is equal to three times the perimeter.

Problem 7.2 Let a, b, c be nonzero integers, $a \neq c$, such that

$$\frac{a}{c} = \frac{a^2 + b^2}{c^2 + d^2}.$$

Prove that $a^2 + b^2 + c^2$ cannot be a prime number.

Problem 7.3 Let $ABCD$ be a convex quadrilateral with $\angle BAC = \angle CAD$ and $\angle ABC = \angle ACD$. Rays AD and BC meet at E and rays AB and DC meet at F. Prove that

(a) $AB \cdot DE = BC \cdot CE$;

(b) $AC^2 < (AD \cdot AF + AB \cdot AE)/2$.

Problem 7.4 In triangle ABC, D and E lie on BC and AB, respectively, F lies on AC such that $EF \parallel BC$, G lies on BC such that $EG \parallel AD$. Let M and N be the midpoints of AD and BC, respectively. Prove that

(a) $\dfrac{EF}{BC} + \dfrac{EG}{AD} = 1$;

(b) the midpoint of FG lies on line MN.

Problem 8.1 Let $p(x) = 2x^3 - 3x^2 + 2$, and let

$$S = \{p(n) \mid n \in \mathbb{N}, n \leq 1999\},$$
$$T = \{n^2 + 1 \mid n \in \mathbb{N}\},$$
$$U = \{n^2 + 2 \mid n \in \mathbb{N}\}.$$

Prove that $S \cap T$ and $S \cap U$ have the same number of elements.

Problem 8.2

(a) Let $n \geq 2$ be a positive integer and

$$x_1, y_1, x_2, y_2, \cdots, x_n, y_n$$

be positive real numbers such that

$$x_1 + x_2 + \cdots + x_n \geq x_1y_1 + x_2y_2 + \cdots + x_ny_n,$$

prove that

$$x_1 + x_2 + \cdots + x_n \leq \frac{x_1}{y_1} + \frac{x_2}{y_2} + \cdots + \frac{x_n}{y_n}.$$

(b) Let a, b, c be positive real numbers such that

$$ab + bc + ca \leq 3abc.$$

Prove that

$$a^3 + b^3 + c^3 \geq a + b + c.$$

Problem 8.3 Let $ABCDA'B'C'D'$ be a rectangular box, let E and F be the feet of perpendiculars from A to lines $A'D$ and $A'C$ respectively, and let P and Q be the feet of perpendiculars from B' to lines $A'C'$ and $A'C$ respectively. Prove that

(a) planes AEF and $B'PQ$ are parallel;

(b) triangles AEF and $B'PQ$ are similar.

Problem 8.4 Let $SABC$ be a right pyramid with equilateral base ABC, let O be the center of ABC, and let M be the midpoint of BC. If $AM = 2SO$ and N is a point on SA such that $SA = 25SN$, prove that planes ABP and SBC are perpendicular, where P is the intersection of SO and MN.

Problem 9.1 Let ABC be a triangle with angle bisector AD. One considers the points M, N on rays AB and AC respectively, such that $\angle MDA = \angle ABC$ and $\angle NDA = \angle BCA$. Lines AD and MN meet at P. Prove that

$$AD^3 = AB \cdot AC \cdot AP.$$

Problem 9.2 For $a, b > 0$, denote by $t(a, b)$ the positive root of the equation

$$(a + b)x^2 - 2(ab - 1)x - (a + b) = 0.$$

Let $M = \{(a, b) \mid a \neq b, t(a, b) \leq \sqrt{ab}\}$. Determine, for $(a, b) \in M$, the minimum value of $t(a, b)$.

Problem 9.3 In the convex quadrilateral $ABCD$ the bisectors of angles A and C meet at I. Prove that $ABCD$ is cyclic if and only if

$$[AIB] + [CID] = [AID] + [BIC].$$

Problem 9.4

(a) Let $a, b \in \mathbb{R}$, $a < b$. Prove that $a < x < b$ if and only if there exists $0 < \lambda < 1$ such that $x = \lambda a + (1 - \lambda)b$.

(b) The function $f : \mathbb{R} \to \mathbb{R}$ has the property:

$$f(\lambda x + (1 - \lambda)y) < \lambda f(x) + (1 - \lambda)f(y)$$

for all $x, y \in \mathbb{R}$, $x \neq y$, and all $0 < \lambda < 1$. Prove that one cannot find four points on the function's graph that are the vertices of a parallelogram.

Problem 10.1 Find all real numbers x and y satisfying

$$\frac{1}{4^x} + \frac{1}{27^y} = \frac{5}{6}$$
$$\log_{27} y - \log_4 x \geq \frac{1}{6}$$
$$27^y - 4^x \leq 1.$$

Problem 10.2 Points M, N, P, Q are on edges AB, BC, CD, DA of the regular tetrahedron $ABCD$, respectively. Prove that

$$MN \cdot NP \cdot PQ \cdot QM \geq AM \cdot BN \cdot CP \cdot DQ.$$

Problem 10.3 Let a, b, c $(c \neq 0)$ be complex numbers. Let z_1 and z_2 be the roots of the equation $az^2 + bz + c = 0$, and let w_1 and w_2 be the roots of the equation

$$(a + \overline{c})z^2 + (b + \overline{b})z + (\overline{a} + c) = 0.$$

Prove that if $|z_1|, |z_2| < 1$, then $|w_1| = |w_2| = 1$.

Problem 10.4

(a) Let $x_1, y_1, x_2, y_2, \ldots, x_n, y_n$ be positive real numbers such that

(i) $x_1 y_1 < x_2 y_2 < \cdots < x_n y_n$;

(ii) $x_1 + x_2 + \cdots + x_k \geq y_1 + y_2 + \cdots + y_k$ for all $k = 1, 2, \ldots, n$.

Prove that

$$\frac{1}{x_1} + \frac{1}{x_2} + \cdots + \frac{1}{x_n} \leq \frac{1}{y_1} + \frac{1}{y_2} + \cdots + \frac{1}{y_n}.$$

(b) Let $A = \{a_1, a_2, \ldots, a_n\} \subset \mathbb{N}$ be a set such that for all distinct subsets $B, C \subseteq A$, $\sum_{x \in B} x \neq \sum_{x \in C} x$. Prove that

$$\frac{1}{a_1} + \frac{1}{a_2} + \cdots + \frac{1}{a_n} < 2.$$

IMO Selection Tests

Problem 1

(a) Show that out of any 39 consecutive positive integers, it is possible to choose one number with the sum of its digits divisible by 11.

(b) Find the first 38 consecutive positive integers, none with the sum of its digits divisible by 11.

Problem 2 Let ABC be an acute triangle with angle bisectors BL and CM. Prove that $\angle A = 60^\circ$ if and only if there exists a point K on BC ($K \neq B, C$) such that triangle KLM is equilateral.

Problem 3 Show that for any positive integer n, the number

$$S_n = \binom{2n+1}{0} \cdot 2^{2n} + \binom{2n+1}{2} \cdot 2^{2n-2} \cdot 3 + \cdots + \binom{2n+1}{2n} \cdot 3^n$$

is the sum of two consecutive perfect squares.

Problem 4 Show that for all positive real numbers $x_1, x_2, \cdots, x_n$ such that

$$x_1 x_2 \cdots x_n = 1,$$

the following inequality holds:

$$\frac{1}{n-1+x_1} + \frac{1}{n-1+x_2} + \cdots + \frac{1}{n-1+x_n} \leq 1.$$

Problem 5 Let $x_1, x_2, \ldots, x_n$ be distinct positive integers. Prove that

$$x_1^2 + x_2^2 + \cdots + x_n^2 \geq \frac{(2n+1)(x_1 + x_2 + \cdots + x_n)}{3}.$$

Problem 6 Prove that for any integer n, $n \geq 3$, there exist n positive integers $a_1, a_2, \ldots, a_n$ in arithmetic progression, and n positive integers $b_1, b_2, \ldots, b_n$ in geometric progression, such that

$$b_1 < a_1 < b_2 < a_2 < \cdots < b_n < a_n.$$

Give an example of two such progressions each having at least 5 terms.

Problem 7 Let a be a positive real number and $\{x_n\}$ $(n \geq 1)$ be a sequence of real numbers such that $x_1 = a$ and

$$x_{n+1} \geq (n+2)x_n - \sum_{k=1}^{n-1} kx_k,$$

for all $n \geq 1$. Show that there exists a positive integer n such that $x_n > 1999!$.

Problem 8 Let O, A, B, C be variable points in the plane such that $OA = 4, OB = 2\sqrt{3}$ and $OC = \sqrt{22}$. Find the maximum possible area of triangle ABC.

Problem 9 Let a, n be integers and let p be a prime such that $p > |a|+1$. Prove that the polynomial $f(x) = x^n + ax + p$ can not be represented as a product of two nonconstant polynomials with integer coefficients.

Problem 10 Two circles meet at A and B. Line ℓ passes through A and meets the circles again at C and D respectively. Let M and N be the midpoints of arcs $\overset{\frown}{BC}$ and $\overset{\frown}{BD}$, which do not contain A, and let K be the midpoint of CD. Prove that $\angle MKN = 90°$.

Problem 11 Let $n \geq 3$ and $A_1, A_2, \ldots, A_n$ be points on a circle. Find the greatest number of acute triangles having vertices in these points.

Problem 12 The scientists at an international conference are either *native* or *foreign*. Each native scientist sends exactly one message to a foreign scientist and each foreign scientist sends exactly one message to a native scientist, although at least one native scientist does not receive a message. Prove that there exists a set S of native scientists such that the native scientists not in S are exactly those who received messages from those foreign scientists who did not receive messages from scientists belonging to S.

Problem 13 A polyhedron P is given in space. Determine whether there must exist three edges of P that can be the sides of a triangle.

3.16 Russia

Fourth round

Problem 8.1 A father wishes to take his two sons to visit their grandmother, who lives 33 kilometers away. He owns a motorcycle whose maximum speed is 25 km/h. With one passenger, its maximum speed drops to 20 km/h. (He cannot carry two passengers.) Each brother walks at a speed of 5 km/h. Show that all three of them can reach the grandmother's house in 3 hours.

Problem 8.2 The natural number A has the following property: the sum of the integers from 1 to A, inclusive, has decimal expansion equal to that of A followed by three digits. Find A.

Problem 8.3 On sides BC, CA, AB of triangle ABC lie points A_1, B_1, C_1 such that the medians A_1A_2, B_1B_2, C_1C_2 of triangle $A_1B_1C_1$ are parallel to AB, BC, CA, respectively. Determine in what ratios the points A_1, B_1, C_1 divide the sides of ABC.

Problem 8.4 We are given 40 balloons, the air pressure inside each of which is unknown and may differ from balloon to balloon. It is permitted to choose up to k of the balloons and equalize the pressure in them (to the arithmetic mean of their respective original pressures). What is the smallest k for which it is always possible to equalize the pressures in all of the balloons?

Problem 8.5 Show that the numbers from 1 to 15 cannot be divided into a group A of 2 numbers and a group B of 13 numbers in such a way that the sum of the numbers in B is equal to the product of the numbers in A.

Problem 8.6 Given an acute triangle ABC, let A_1 be the reflection of A across the line BC, and let C_1 be the reflection of C across the line AB. Show that if A_1, B, C_1 lie on a line and $C_1B = 2A_1B$, then $\angle CA_1B$ is a right angle.

Problem 8.7 In a box lies a complete set of 1×2 dominoes. (That is, for each pair of integers i, j with $0 \le i \le j \le n$, there is one domino with i on one square and j on the other.) Two players take turns selecting one domino from the box and adding it to one end of an open (straight) chain

---

[1] Problems are numbered as they appeared in the contests. Repetition problems are skipped.

on the table, so that adjacent dominoes have the same numbers on their adjacent squares. (The first player's move may be any domino.) The first player unable to move loses. Which player wins with correct play?

Problem 8.8 An open chain of 54 squares of side length 1 is made so that each pair of consecutive squares is joined at a single vertex, and each square is joined to its two neighbors at opposite vertices. Is it possible to cover the surface of a $3 \times 3 \times 3$ cube with this chain?

Problem 9.1 Around a circle are written all of the positive integers from 1 to N, $N \geq 2$, in such a way that any two adjacent integers have at least one common digit in their decimal expansions. Find the smallest N for which this is possible.

Problem 9.2 In triangle ABC, points D and E are chosen on side CA such that $AB = AD$ and $BE = EC$ (E lying between A and D). Let F be the midpoint of the arc BC of the circumcircle of ABC. Show that B, E, D, F lie on a circle.

Problem 9.3 The product of the positive real numbers x, y, z is 1. Show that if

$$1/x + 1/y + 1/z \geq x + y + z,$$

then

$$\frac{1}{x^k} + \frac{1}{y^k} + \frac{1}{z^k} \geq x^k + y^k + z^k$$

for all positive integers k.

Problem 9.4 A maze consists of an 8×8 grid, in each 1×1 cell of which is drawn an arrow pointing up, down, left or right. The top edge of the top right square is the exit from the maze. A token is placed on the bottom left square, and then is moved in a sequence of turns. On each turn, the token is moved one square in the direction of the arrow. Then the arrow in the square the token moved from is rotated $90°$ clockwise. If the arrow points off of the board (and not through the exit), the token stays put and the arrow is rotated $90°$ clockwise. Prove that sooner or later the token will leave the maze.

Problem 9.5 Each square of an infinite grid is colored in one of 5 colors, in such a way that every 5-square (Greek) cross contains one square of each color. Show that every 1×5 rectangle also contains one square of each color.

Problem 9.7 Show that each natural number can be written as the difference of two natural numbers having the same number of prime factors.

Problem 9.8 In triangle ABC, with $AB > BC$, points K and M are the midpoints of sides AB and CA, and I is the incenter. Let P be the intersection of the lines KM and CI, and Q the point such that $QP \perp KM$ and $QM \parallel BI$. Prove that $QI \perp AC$.

Problem 10.2 In the plane is given a circle ω, a point A inside ω, and a point B not equal to A. Consider all possible triangles BXY such that X and Y lie on ω and A lies on the chord XY. Show that the circumcenters of these triangles all lie on a line.

Problem 10.3 In space are given n points in general position (no three points are collinear and no four are coplanar). Through any three of them is drawn a plane. Show that for any $n - 3$ points in space, there exists one of the drawn planes not passing through any of these points.

Problem 10.5 Do there exist 10 distinct integers, the sum of any 9 of which is a perfect square?

Problem 10.6 The incircle of triangle ABC touches sides BC, CA, AB at A_1, B_1, C_1, respectively. Let K be the point on the circle diametrically opposite C_1, and D the intersection of the lines B_1C_1 and A_1K. Prove that $CD = CB_1$.

Problem 10.7 Each voter in an election marks on a ballot the names of n candidates. Each ballot is placed into one of $n + 1$ boxes. After the election, it is observed that each box contains at least one ballot, and that for any $n+1$ ballots, one in each box, there exists a name which is marked on all of these ballots. Show that for at least one box, there exists a name which is marked on all ballots in the box.

Problem 10.8 A set of natural numbers is chosen so that among any 1999 consecutive natural numbers, there is a chosen number. Show that there exist two chosen numbers, one of which divides the other.

Problem 11.1 The function $f(x)$ is defined on all real numbers. It is known that for all $a > 1$, the function $f(x) + f(ax)$ is continuous. Show that $f(x)$ is continuous.

Problem 11.3 In a class, each boy is friends with at least one girl. Show that there exists a group of at least half of the students, such that each boy in the group is friends with an odd number of the girls in the group.

Problem 11.4 A polyhedron is circumscribed about a sphere. We call a face big if the projection of the sphere onto the plane of the face lies entirely within the face. Show that there are at most 6 big faces.

Problem 11.5 Do there exist real numbers a, b, c such that for all real numbers x, y,

$$|x+a|+|x+y+b|+|y+c|>|x|+|x+y|+|y|?$$

Problem 11.6 Each cell of a 50×50 square is colored in one of four colors. Show that there exists a cell which has cells of the same color directly above, directly below, directly to the left, and directly to the right of it (though not necessarily adjacent to it).

Problem 11.8 A polynomial with integer coefficients has the property that there exist infinitely many integers which are the value of the polynomial evaluated at more than one integer. Prove that there exists at most one integer which is the value of the polynomial at exactly one integer.

Fifth round

Problem 9.1 In the decimal expansion of A, the digits occur in increasing order from left to right. What is the sum of the digits of $9A$?

Problem 9.2 In a country are several cities, some pairs of which are connected by a direct flight on one of N airlines, in such a way that each airline has at least one flight to each city. It is known that any city can be reached from any other city by a series of flights. A financial crisis forces the cancellation of $N-1$ flights, with no airline losing more than one flight. Show that any city can still be reached from any other city.

Problem 9.3 Let S be the circumcircle of triangle ABC. Let A_0 be the midpoint of the arc BC of S not containing A, and C_0 the midpoint of the arc AB of S not containing C. Let S_1 be the circle with center A_0 tangent to BC, and let S_2 be the circle with center C_0 tangent to AB. Show that the incenter I of ABC lies on a common external tangent to S_1 and S_2.

Problem 9.4 The numbers from 1 to 1000000 can be colored black or white. A permissible move consists of selecting a number from 1 to 1000000 and changing the color of that number and each number not relatively prime to it. Initially all of the numbers are black. Is it possible to make a sequence of moves after which all of the numbers are colored white?

Problem 9.5 An equilateral triangle of side length n is drawn with sides along a triangular grid of side length 1. What is the maximum number of grid segments on or inside the triangle that can be marked so that no three marked segments form a triangle?

Problem 9.6 Let $\{x\} = x - \lfloor x \rfloor$ denote the fractional part of x. Prove that for every natural number n,

$$\sum_{k=1}^{n^2} \{\sqrt{k}\} \leq \frac{n^2 - 1}{2}.$$

Problem 9.7 A circle passing through vertices A and B of triangle ABC intersects side BC again at D. A circle passing through vertices B and C intersects side AB again at E, and intersects the first circle again at F. Suppose that the points A, E, D, C lie on a circle centered at O. Show that $\angle BFO$ is a right angle.

Problem 9.8 A circuit board has 2000 contacts, any two of which are connected by a lead. The hooligans Vasya and Petya take turns cutting leads: Vasya (who goes first) always cuts one lead, while Petya cuts either one or three leads. The first person to cut the last lead from some contact loses. Who wins with correct play?

Problem 10.1 Three empty bowls are placed on a table. Three players A, B, C, whose order of play is determined randomly, take turns putting one token into a bowl. A can place a token in the first or second bowl, B in the second or third bowl, and C in the third or first bowl. The first player to put the 1999th token into a bowl loses. Show that players A and B can work together to ensure that C will lose.

Problem 10.2 Find all infinite bounded sequences $a_1, a_2, \ldots$ of positive integers such that for all $n > 2$,

$$a_n = \frac{a_{n-1} + a_{n-2}}{\gcd(a_{n-1}, a_{n-2})}.$$

Problem 10.3 The incircle of triangle ABC touches sides AB, BC, CA at K, L, M, respectively. For each two of the incircles of AMK, BKL, CLM is drawn the common external tangent not lying along a side of ABC. Show that these three tangents pass through a single point.

Problem 10.4 An $n \times n$ square is drawn on an infinite checkerboard. Each of the n^2 cells contained in the square initially contains a token. A move consists of jumping a token over an adjacent token (horizontally or vertically) into an empty square; the token jumped over is removed. A sequence of moves is carried out in such a way that at the end, no further moves are possible. Show that at least $n^2/3$ moves have been made.

Problem 10.5 The sum of the decimal digits of the natural number n is 100, and that of $44n$ is 800. What is the sum of the digits of $3n$?

Problem 10.7 The positive real numbers x and y satisfy

$$x^2 + y^3 \geq x^3 + y^4.$$

Show that $x^3 + y^3 \leq 2$.

Problem 10.8 In a group of 12 people, among every 9 people one can find 5 people, any two of whom know each other. Show that there exist 6 people in the group, any two of whom know each other.

Problem 11.1 Do there exist 19 distinct natural numbers with the same sum of digits, the sum of which is 1999?

Problem 11.2 At each rational point on the real line is written an integer. Show that there exists a segment with rational endpoints, such that the sum of the numbers at the endpoints does not exceed twice the number at the midpoint.

Problem 11.3 A circle inscribed in quadrilateral $ABCD$ touches sides DA, AB, BC, CD at K, L, M, N, respectively. Let S_1, S_2, S_3, S_4 be the incircles of triangles AKL, BLM, CMN, DNK, respectively. The common external tangents to S_1 and S_2, to S_2 and S_3, to S_3 and S_4, and to S_4 and S_1, not lying on the sides of $ABCD$, are drawn. Show that the quadrilateral formed by these tangents is a rhombus.

Problem 11.5 Four natural numbers have the property that the square of the sum of any two of the numbers is divisible by the product of the other two. Show that at least three of the four numbers are equal.

Problem 11.6 Show that three convex polygons in the plane cannot be intersected by a single line if and only if for each of the polygons, there exists a line intersecting none of the polygons, such that the given polygon lies on the opposite side of the line from the other two.

Problem 11.7 Through vertex A of tetrahedron $ABCD$ passes a plane tangent to the circumscribed sphere of the tetrahedron. Show that the lines of intersection of the plane with the planes ABC, ACD, ABD form six equal angles if and only if

$$AB \cdot CD = AC \cdot BD = AD \cdot BC.$$

3.17 Taiwan

Problem 1 Determine all solutions (x, y, z) of positive integers such that

$$(x+1)^{y+1}+1=(x+2)^{z+1}.$$

Problem 2 There are 1999 people participating in an exhibition. Out of any 50 people, at least 2 do not know each other. Prove that we can find at least 41 people who each know at most 1958 other people.

Problem 3 Let P^* denote all the odd primes less than 10000, and suppose $p \in P^*$. For each subset $S=\{p_1, p_2, \cdots, p_k\}$ of P^*, with $k \geq 2$ and not including p, there exists a $q \in P^* \setminus S$ such that

$$q+1 \mid (p_1+1)(p_2+1)\cdots(p_k+1).$$

Find all such possible values of p.

Problem 4 The altitudes through the vertices A, B, C of an acute-angled triangle ABC meet the opposite sides at D, E, F, respectively, and $AB > AC$. The line EF meets BC at P, and the line through D parallel to EF meets the lines AC and AB at Q and R, respectively. Let N be a point on the side BC such that $\angle NQP + \angle NRP < 180°$. Prove that $BN > CN$.

Problem 5 There are 8 different symbols designed on n different T-shirts, where $n \geq 2$. It is known that each shirt contains at least one symbol, and for any two shirts, the symbols on them are not all the same. Also, for any k symbols, $1 \leq k \leq 7$, the number of shirts containing at least one of the k symbols is even. Find the value of n.

3.18 Turkey

Problem 1 Let ABC be an isosceles triangle with $AB = AC$. Let D be a point on BC such that $BD = 2DC$, and let P be a point on AD such that $\angle BAC = \angle BPD$. Prove that

$$\angle BAC = 2\angle DPC.$$

Problem 2 Prove that

$$(a+3b)(b+4c)(c+2a) \geq 60abc$$

for all real numbers $0 \leq a \leq b \leq c$.

Problem 3 The points on a circle are colored in three different colors. Prove that there exist infinitely many isosceles triangles with vertices on the circle and of the same color.

Problem 4 Let $\angle XOY$ be a given angle, and let M and N be two points on the rays OX and OY, respectively. Determine the locus of the midpoint of MN as M and N varies along the rays OX and OY such that $OM + ON$ is constant.

Problem 5 Some of the vertices of the unit squares of an $n \times n$ chessboard are colored such that any $k \times k$ square formed by these unit squares has a colored point on at least one of its sides. If $l(n)$ denotes the minimum number of colored points required to ensure the above condition, prove that

$$\lim_{n\to\infty} \frac{l(n)}{n^2} = \frac{2}{7}.$$

Problem 6 Let $ABCD$ be a cyclic quadrilateral, and let L and N be the midpoints of diagonals AC and BD, respectively. Suppose that BD bisects $\angle ANC$. Prove that AC bisects $\angle BLD$.

Problem 7 Determine all functions $f : \mathbb{R} \to \mathbb{R}$ such that the set

$$\left\{ \frac{f(x)}{x} \mid x \in \mathbb{R} \text{ and } x \neq 0 \right\}$$

is finite and

$$f(x - 1 - f(x)) = f(x) - x - 1$$

for all $x \in \mathbb{R}$.

Problem 8 Let the area and the perimeter of a cyclic quadrilateral C be A_C and P_C, respectively. If the area and the perimeter of the quadrilateral which is tangent to the circumcircle of C at the vertices of C are A_T and P_T, respectively, prove that

$$\frac{A_C}{A_T} \geq \left(\frac{P_C}{P_T}\right)^2.$$

Problem 9 Prove that the plane is not a union of the inner regions of finitely many parabolas. (The outer region of a parabola is the union of the lines on the plane not intersecting the parabola. The inner region of a parabola is the set of points on the plane that do not belong to the outer region of the parabola.)

3.19 Ukraine

Problem 1 Let $P(x)$ be a polynomial with integer coefficients. The sequence $\{x_n\}_{n\geq 1}$ satisfies the conditions $x_1 = x_{2000} = 1999$, and $x_{n+1} = P(x_n)$ for $n \geq 1$. Calculate

$$\frac{x_1}{x_2} + \frac{x_2}{x_3} + \cdots + \frac{x_{1999}}{x_{2000}}.$$

Problem 2 For real numbers $0 \leq x_1, x_2, \ldots, x_6 \leq 1$ prove the inequality

$$\frac{x_1^3}{x_2^5 + x_3^5 + x_4^5 + x_5^5 + x_6^5 + 5} + \frac{x_2^3}{x_1^5 + x_3^5 + x_4^5 + x_5^5 + x_6^5 + 5} + \cdots + \frac{x_6^3}{x_1^5 + x_2^5 + x_3^5 + x_4^5 + x_5^5 + 5} \leq \frac{3}{5}.$$

Problem 3 Let AA_1, BB_1, CC_1 be the altitudes of an acute triangle ABC, and let O be an arbitrary point inside the triangle $A_1B_1C_1$. Let M, N, P, Q, R, S be the orthogonal projections of O onto lines AA_1, BC, BB_1, CA, CC_1, AB, respectively. Prove that lines MN, PQ, RS are concurrent.

3.20 United Kingdom

Problem 1 I have four children. The age in years of each child is a positive integer between 2 and 16 inclusive and all four ages are distinct. A year ago the square of the age of the oldest child was equal to sum of the squares of the ages of the other three. In one year's time the sum of the squares of the ages of the oldest and the youngest children will be equal to the sum of the squares of the other two children. Decide whether this information is sufficient to determine their ages uniquely, and find all possibilities for their ages.

Problem 2 A circle has diameter AB and X is a fixed point on the segment AB. A point P, distinct from A and B, lies on the circle. Prove that, for all possible positions of P,

$$\frac{\tan \angle APX}{\tan \angle PAX}$$

is a constant.

Problem 3 Determine a positive constant c such that the equation

$$xy^2 - y^2 - x + y = c$$

has exactly three solutions (x, y) in positive integers.

Problem 4 Any positive integer m can be written uniquely in base 3 form as a string of 0's, 1's and 2's (not beginning with a zero). For example,

$$98 = 81 + 9 + 2 \times 3 + 2 \times 1 = (10122)_3.$$

Let $c(m)$ denote the sum of the cubes of the digits of the base 3 form of m; thus, for instance

$$c(98) = 1^3 + 0^3 + 1^3 + 2^3 + 2^3 = 18.$$

Let n be any fixed positive integer. Define the sequence $\{u_r\}$ as

$$u_1 = n, \text{ and } u_r = c(u_{r-1}) \text{ for } r \geq 2.$$

Show that there is a positive integer r such that $u_r = 1, 2$, or 17.

Problem 5 Consider all functions $f : \mathbb{N} \to \mathbb{N}$ such that

(i) for each positive integer m, there is a unique positive integer n such that $f(n) = m$;

(ii) for each positive integer n, $f(n+1)$ is either $4f(n)-1$ or $f(n)-1$.

Find the set of positive integers p such that $f(1999) = p$ for some function f with properties (i) and (ii).

Problem 6 For each positive integer n, let $S_n = \{1, 2, \ldots, n\}$.

(a) For which values of n is it possible to express S_n as the union of two non-empty disjoint subsets so that the elements in the two subsets have equal sum?

(b) For which values of n is it possible to express S_n as the union of three non-empty disjoint subsets so that the elements in the three subsets have equal sum?

Problem 7 Let $ABCDEF$ be a hexagon which circumscribes a circle ω. The circle ω touches AB, CD, EF at their respective midpoints P, Q, R. Let ω touch BC, DE, FA at X, Y, Z respectively. Prove that PY, QZ, RX are concurrent.

Problem 8 Some three non-negative real numbers p, q, r satisfy

$$p + q + r = 1.$$

Prove that

$$7(pq + qr + rp) \leq 2 + 9pqr.$$

Problem 9 Consider all numbers of the form $3n^2 + n + 1$, where n is a positive integer.

(a) How small can the sum of the digits (in base 10) of such a number be?

(b) Can such a number have the sum of its digits (in base 10) equal to 1999?

3.21 United States of America

Problem 1 Some checkers placed on an $n \times n$ checkerboard satisfy the following conditions:

(i) every square that does not contain a checker shares a side with one that does;

(ii) given any pair of squares that contain checkers, there is a sequence of squares containing checkers, starting and ending with the given squares, such that every two consecutive squares of the sequence share a side.

Prove that at least $(n^2 - 2)/3$ checkers have been placed on the board.

Problem 2 Let $ABCD$ be a cyclic quadrilateral. Prove that

$$|AB - CD| + |AD - BC| \geq 2|AC - BD|.$$

Problem 3 Let $p > 2$ be a prime and let a, b, c, d be integers not divisible by p, such that

$$\{ra/p\} + \{rb/p\} + \{rc/p\} + \{rd/p\} = 2$$

for any integer r not divisible by p. Prove that at least two of the numbers $a+b$, $a+c$, $a+d$, $b+c$, $b+d$, $c+d$ are divisible by p. Here, for real number x, $\{x\} = x - \lfloor x \rfloor$ denotes the fractional part of x.

Problem 4 Let $a_1, a_2, \ldots, a_n$ $(n > 3)$ be real numbers such that

$$a_1 + a_2 + \cdots + a_n \geq n \qquad \text{and} \qquad a_1^2 + a_2^2 + \cdots + a_n^2 \geq n^2.$$

Prove that $\max(a_1, a_2, \ldots, a_n) \geq 2$.

Problem 5 The Y2K Game is played on a 1×2000 grid as follows. Two players in turn write either an S or an O in an empty square. The first player who produces three consecutive boxes that spell SOS wins. If all boxes are filled without producing SOS then the game is a draw. Prove that the second player has a winning strategy.

Problem 6 Let $ABCD$ be an isosceles trapezoid with $AB \parallel CD$. The inscribed circle ω of triangle BCD meets CD at E. Let F be a point on the (internal) angle bisector of $\angle DAC$ such that $EF \perp CD$. Let the circumscribed circle of triangle ACF meet line CD at C and G. Prove that the triangle AFG is isosceles.

3.22 Vietnam

Problem 1 Solve the system of equations

$$(1+4^{2x-y})\cdot 5^{1-2x+y} = 1+2^{2x-y+1}$$
$$y^3+4x+1+\ln(y^2+2x)=0.$$

Problem 2 Let A', B', C' be the respective midpoints of the arcs BC, CA, AB, not containing points A, B, C, respectively, of the circumcircle of the triangle ABC. The sides BC, CA, AB meet the pairs of segments

$$\{C'A', A'B'\},\ \{A'B', B'C'\},\ \{B'C', C'A'\}$$

at the pairs of points

$$\{M, N\},\ \{P, Q\},\ \{R, S\},$$

respectively. Prove that $MN = PQ = RS$ if and only if the triangle ABC is equilateral.

Problem 3 For $n = 0, 1, 2, \ldots$, let $\{x_n\}$ and $\{y_n\}$ be two sequences defined recursively as follows:

$$x_0 = 1,\ x_1 = 4,\ x_{n+2} = 3x_{n+1} - x_n;$$

$$y_0 = 1,\ y_1 = 2,\ y_{n+2} = 3y_{n+1} - y_n.$$

(a) Prove that $x_n^2 - 5y_n^2 + 4 = 0$ for all non-negative integers n.

(b) Suppose that a, b are two positive integers such that $a^2 - 5b^2 + 4 = 0$. Prove that there exists a non-negative integer k such that $x_k = a$ and $y_k = b$.

Problem 4 Let a, b, c be real numbers such that $abc + a + c = b$. Determine the greatest possible value of the expression

$$P = \frac{2}{a^2+1} - \frac{2}{b^2+1} + \frac{3}{c^2+1}.$$

Problem 5 In the three-dimensional space let Ox, Oy, Oz, Ot be four nonplanar distinct rays such that the angles between any two of them have the same measure.

(a) Determine this common measure.

(b) Let Or be another ray different from the above four rays. let $\alpha, \beta, \gamma, \delta$ be the angles formed by Or with Ox, Oy, Oz, Ot, respectively. Put

$$p = \cos\alpha + \cos\beta + \cos\gamma + \cos\delta,$$

$$q = \cos^2\alpha + \cos^2\beta + \cos^2\gamma + \cos^2\delta.$$

Prove that p and q are invariables when Or rotates about the point O.

Problem 6 Let $\mathcal{S} = \{0, 1, 2, \ldots, 1999\}$ and $\mathcal{T} = \{0, 1, 2, \ldots\}$. Find all functions $f : \mathcal{T} \to \mathcal{S}$ such that

(i) $f(s) = s$ for all $s \in \mathcal{S}$.

(ii) $f(m+n) = f(f(m) + f(n))$ for all $m, n \in \mathcal{T}$.

Problem 7 For $n = 1, 2, \ldots$, let $\{u_n\}$ be a sequence defined by

$$u_1 = 1,\ u_2 = 2,\ u_{n+2} = 3u_{n+1} - u_n.$$

Prove that

$$u_{n+2} + u_n \geq 2 + \frac{u_{n+1}^2}{u_n}$$

for all n.

Problem 8 Let ABC be a triangle inscribed in circle ω. Locate the position of the points P, not lying in ω, in the plane (ABC) with the property that the lines PA, PB, PC meet ω again at points A', B', C' such that $A'B' \perp A'C'$ and $A'B' = A'C'$.

Problem 9 Consider real numbers a, b such that all roots of the equation

$$ax^3 - x^2 + bx - 1 = 0$$

are real and positive. Determine the smallest possible value of the expression

$$P = \frac{5a^2 - 3ab + 2}{a^2(b-a)}.$$

Problem 10 Let $f(x)$ be a continuous function defined on $[0, 1]$ such that

(i) $f(0) = f(1) = 0$;

(ii) $2f(x) + f(y) = 3f\left(\frac{2x+y}{3}\right)$ for all $x, y \in [0, 1]$.

Prove that $f(x) = 0$ for all $x \in [0, 1]$.

Problem 11 The base side and the altitude of a regular hexagonal prism $ABCDEF - A'B'C'D'E'F'$ are equal to a and h respectively.

(a) Prove that six planes

$$(AB'F), (CD'B'), (EF'D'), (D'EC), (F'AE), (B'CA)$$

touch the same sphere.

(b) Determine the center and the radius of the sphere.

Problem 12 For $n = 1, 2, \ldots$, two sequences $\{x_n\}$ and $\{y_n\}$ are defined recursively by

$$x_1 = 1,\ y_1 = 2,\ x_{n+1} = 22y_n - 15x_n,\ y_{n+1} = 17y_n - 12x_n.$$

(a) Prove that x_n and y_n are not equal to zero for all $n = 1, 2, \ldots$.

(b) Prove that each sequence contains infinitely many positive terms and infinitely many negative terms.

(c) For $n = 1999^{1945}$, determine whether x_n and y_n are divisible by 7 or not.

4

1999 Regional Contests: Problems

4.1 Asian Pacific Mathematical Olympiad

Problem 1 Find the smallest positive integer n with the following property: There does not exist an arithmetic progression of 1999 real numbers containing exactly n integers.

Problem 2 Let a_1, a_2, ... be a sequence of real numbers satisfying

$$a_{i+j} \leq a_i + a_j$$

for all i, j = 1, 2, ... Prove that

$$a_1 + \frac{a_2}{2} + \frac{a_3}{3} + \cdots + \frac{a_n}{n} \geq a_n$$

for all positive integers n.

Problem 3 Let ω_1 and ω_2 be two circles intersecting at P and Q. The common tangent, closer to P, of ω_1 and ω_2, touches ω_1 at A and ω_2 at B. The tangent to ω_1 at P meets ω_2 again at C, and the extension of AP meets BC at R. Prove that the circumcircle of triangle PQR is tangent to BP and BR.

Problem 4 Determine all pairs (a, b) of integers for which the numbers $a^2 + 4b$ and $b^2 + 4a$ are both perfect squares.

Problem 5 Let S be a set of $2n + 1$ points in the plane such that no three are collinear and no four concyclic. A circle will be called *good* if it has 3 points of S on its circumference, $n - 1$ points in its interior, and $n - 1$ in its exterior. Prove that the number of good circles has the same parity as n.

4.2 Austrian-Polish Mathematics Competition

Problem 1 Let n be a positive integer and $M = \{1, 2, \ldots, n\}$. Find the number of ordered 6-tuples $(A_1, A_2, A_3, A_4, A_5, A_6)$ which satisfy the following two conditions:

(a) A_1, A_2, A_3, A_4, A_5, A_6 (not necessarily distinct) are subsets of M;

(b) each element of M belongs to either 0, 3, or 6 of the sets A_1, A_2, A_3, A_4, A_5, A_6.

Problem 2 Find the largest real number C_1 and the smallest real number C_2 such that for all positive real numbers a, b, c, d, e the following inequalities hold:

$$C_1 < \frac{a}{a+b} + \frac{b}{b+c} + \frac{c}{c+d} + \frac{d}{d+e} + \frac{e}{e+a} < C_2.$$

Problem 3 Let $n \geq 2$ be a given integer. Determine all systems of n functions $(f_1, \ldots, f_n)$ where $f_i : \mathbb{R} \to \mathbb{R}$, $i = 1, 2, \ldots, n$, such that for all x, $y \in \mathbb{R}$ the following equalities hold:

$$\begin{aligned}
f_1(x) - f_2(x)f_2(y) + f_1(y) &= 0 \\
f_2(x^2) - f_3(x)f_3(y) + f_2(y^2) &= 0 \\
&\vdots \\
f_{n-1}(x^{n-1}) - f_n(x)f_n(y) + f_{n-1}(y^{n-1}) &= 0 \\
f_n(x^n) - f_1(x)f_1(y) + f_n(y^n) &= 0.
\end{aligned}$$

Problem 4 Three straight lines k, l, and m are drawn through an arbitrary point P inside a triangle ABC such that:

(a) k meets the lines AB and AC in A_1 and A_2 ($A_1 \neq A_2$) respectively and $PA_1 = PA_2$;

(b) l meets the lines BC and BA in B_1 and B_2 ($B_1 \neq B_2$) respectively and $PB_1 = PB_2$;

(c) m meets the lines CA and CB in C_1 and C_2 ($C_1 \neq C_2$) respectively and $PC_1 = PC_2$.

Prove that the lines k, l, m are uniquely determined by the above conditions. Find the point P (and prove that there exists exactly one such point) for which the triangles AA_1A_2, BB_1B_2, and CC_1C_2 have the same area.

Problem 5 A sequence of integers $\{a_n\}_{n\geq 1}$ satisfies the following recursive relation

$$a_{n+1} = a_n^3 + 1999 \quad \text{for } n = 1, 2, \ldots.$$

Prove that there exists at most one n for which a_n is a perfect square.

Problem 6 Solve the system of equations

$$\begin{aligned} x_1^2 + x_1x_0 + x_0^4 &= 1 \\ x_2^2 + x_2x_1 + x_1^4 &= 1 \\ x_3^2 + x_3x_2 + x_2^4 &= 1 \\ &\cdots\cdots\cdots \\ x_{1999}^2 + x_{1999}x_{1998} + x_{1998}^4 &= 1 \\ x_0 &= x_{1999} \end{aligned}$$

in the set of nonnegative real numbers.

Problem 7 Find all pairs (x, y) of positive integers such that

$$x^{x+y} = y^{y-x}.$$

Problem 8 Let ℓ be a given straight line and let the points P and Q lie on the same side of the line ℓ. The points M, N lie on the line ℓ and satisfy $PM \perp \ell$ and $QN \perp \ell$. The point S lies between the lines PM and QN such that $PM = PS$ and $QN = QS$. The perpendicular bisectors of SM and SN meet at R. Let T be the second intersection of the line RS and the circumcircle of triangle PQR. Prove that $RS = ST$.

Problem 9 Consider the following one player game. On the plane, a finite set of selected lattice points and segments is called a *position* in this game if the following hold:

(i) the endpoints of each selected segment are lattice points;

(ii) each selected segment is parallel to a coordinate axis, or to the line $y = x$, or to the line $y = -x$;

(iii) each selected segment contains exactly 5 lattice points and all of them are selected;

(iv) any two selected segments have at most one common point.

A move in this game consists of selecting a new lattice point and a new segment such that the new set of selected lattice points and segments is a position. Prove or disprove that there exists an initial position such that the game has infinitely many moves.

4.3 Balkan Mathematical Olympiad

Problem 1 Given an acute-angled triangle ABC, let D be the midpoint of minor arc $\widehat{BC}$ of circumcircle ABC. Let E and F be the respective images of D under reflections about BC and the center of the circumcircle. Finally, let K be the midpoint of AE. Prove that:

(a) the circle passing through the midpoints of the sides of the triangle ABC also passes through K.

(b) the line passing through K and the midpoint of BC is perpendicular to AF.

Problem 2 Let $p > 2$ be a prime number such that $3|(p-2)$. Let

$$\mathcal{S} = \{y^2 - x^3 - 1 \mid x \text{ and } y \text{ are integers, } 0 \leq x, y \leq p-1\}.$$

Prove that at most $p-1$ elements of $\mathcal{S}$ are divisible by p.

Problem 3 Let ABC be an acute triangle, and let M, N, and P be the feet of the perpendiculars from the centroid to the three sides. Prove that

$$\frac{4}{27} < \frac{[MNP]}{[ABC]} \leq \frac{1}{4}.$$

Problem 4 Let $\{x_n\}_{n\geq 0}$ be a nondecreasing sequence of nonnegative integers such that for every $k \geq 0$ the number of terms of the sequence which are less than or equal to k is finite; let this number be y_k. Prove that for all positive integers m and n,

$$\sum_{i=0}^{n} x_i + \sum_{j=0}^{n} y_j \geq (n+1)(m+1).$$

4.4 Czech and Slovak Match

Problem 1 For arbitrary positive numbers a, b, c, prove the inequality

$$\frac{a}{b+2c}+\frac{b}{c+2a}+\frac{c}{a+2b}\geq 1.$$

Problem 2 Let ABC be an acute triangle with altitudes AD, BE, and CF. Let l be the line through D parallel to EF. Let $P = BC \cap EF$, $Q = l \cap AC$, and $R = l \cap AB$. Prove that the circumcircle of triangle PQR passes through the midpoint of BC.

Problem 3 Find all positive integers k for which there exists a ten-element set M of positive numbers such that there are exactly k different triangles whose side lengths are three (not necessarily distinct) elements of M. (Triangles are considered different if they are not congruent.)

Problem 4 Find all positive integers k for which the following assertion holds: If $F(x)$ is a polynomial with integer coefficients which satisfies $F(c) \leq k$ for all $c \in \{0, 1, \ldots, k+1\}$, then

$$F(0) = F(1) = \cdots = F(k+1).$$

Problem 5 Find all functions $f : (1, \infty) \to \mathbb{R}$ such that

$$f(x) - f(y) = (y - x)f(xy)$$

for all $x, y > 1$.

Problem 6 Show that for any positive integer $n \geq 3$, the least common multiple of the numbers 1, 2, ..., n is greater than 2^{n-1}.

4.5 Hong Kong (China)

Problem 1 Let $PQRS$ be a cyclic quadrilateral with $\angle PSR = 90°$, and let H and K be the respective feet of perpendiculars from Q to PR and PS. Prove that HK bisects QS.

Problem 2 The base of a pyramid is a convex nonagon. Each base diagonal and each lateral edge is colored either black or white. Both colors are used at least once. (Note that the sides of the base are not colored.) Prove that there are three segments colored the same color which form a triangle.

Problem 3 Let s and t be nonzero integers, and let (x, y) be any ordered pair of integers. A move changes (x, y) to $(x - t, y - s)$. The pair (x, y) is *good* if after some (possibly zero) number of moves it becomes a pair of integers that are not relatively prime.

(a) Determine if (s, t) is a good pair;

(b) Prove that for any s, t there exists a pair (x, y) which is not good.

Problem 4 Let f be a function defined on the positive reals with the following properties:

(1) $f(1) = 1$;

(2) $f(x + 1) = xf(x)$;

(3) $f(x) = 10^{g(x)}$,

where $g(x)$ is a function defined on the reals satisfying

$$g(ty + (1 - t)z) \leq tg(y) + (1 - t)g(z)$$

for all real y, z and some $0 \leq t \leq 1$.

(a) Prove that

$$t[g(n) - g(n - 1)] \leq g(n + t) - g(n) \leq t[g(n + 1) - g(n)]$$

where n is an integer and $0 \leq t \leq 1$.

(b) Prove that

$$\frac{4}{3} \leq f\left(\frac{1}{2}\right) \leq \frac{4\sqrt{2}}{3}.$$

4.6 Iberoamerican Mathematical Olympiad

Problem 1 Find all positive integers n less than 1000 such that n^2 is equal to the cube of the sum of its digits.

Problem 2 Given two circles ω_1 and ω_2, we say that ω_1 bisects ω_2 if their common chord is a diameter of ω_2. Consider two non-concentric fixed circles ω_1 and ω_2.

(a) Show that there are infinitely many circles ω that bisect both ω_1 and ω_2.

(b) Find the locus of the center of ω.

Problem 3 Let $P_1, P_2, \ldots, P_n$ $(n \geq 2)$ be n distinct collinear points. Circles with diameter P_iP_j $(1 \leq i < j \leq n)$ are drawn and each circle is colored in one of k given colors. All points that belong to more than one circle are not colored. Such a configuration is called a *(n,k)−cover.* For any given k, find all n such that for any (n,k)−cover there exist two lines externally tangent to two circles of the same color.

Problem 4 Let n be an integer greater than 10 such that each of its digits belongs to the set $S = \{1, 3, 7, 9\}$. Prove that n has some prime divisor greater than or equal to 11.

Problem 5 Let ABC be an acute triangle with circumcircle ω centered at O. Let AD, BE, and CF be the altitudes of ABC. Let EF meet ω at P and Q.

(a) Prove that $AO \perp PQ$.

(b) If M is the midpoint of BC, prove that

$$AP^2 = 2AD \cdot OM.$$

Problem 6 Let AB be a segment and C a point on its perpendicular bisector. Construct $C_1, C_2, \ldots, C_n, \ldots$ as follows: $C_1 = C$, and for $n \geq 1$, if C_n is not on AB, then C_{n+1} is the circumcenter of triangle ABC_n. Find all points C such that the sequence $\{C_n\}_{n\geq 1}$ is well defined for all n and such that the sequence eventually becomes periodic.

4.7 Olimpiada de mayo

Problem 1 Find the smallest positive integer n such that the 73 fractions

$$\frac{19}{n+21}, \frac{20}{n+22}, \frac{21}{n+23}, \ldots, \frac{91}{n+93}$$

are all irreducible.

Problem 2 Let ABC be a triangle with $\angle A = 90°$. Construct point P on BC such that if Q is the foot of the perpendicular from P to AC then $PQ^2 = PB \cdot PC$.

Problem 3 There are 1999 balls in a row. Each ball is colored either red or blue. Underneath each ball we write a number equal to the sum of the number of red balls to its right and blue balls to its left. Exactly three numbers each appear on an odd number of balls; determine these three numbers.

Problem 4 Let A be a number with six digits, three of which are colored and are equal to $1, 2, 4$. Prove that it is always possible to obtain a multiple of 7 by doing one of the following:

(1) eliminate the three colored numbers;

(2) write the digits of A in a different order.

Problem 5 Consider a square of side length 1. Let S be a set of finitely many points on the sides of the square. Prove that there is a vertex of the square such that the arithmetic mean of the squares of the distances from the vertex to all the points in S is no less than $3/4$.

Problem 6 An ant crosses a circular disc of radius r and it advances in a straight line, but sometimes it stops. Whenever it stops, it turns $60°$, each time in the opposite direction. (If the last time it turned $60°$ clockwise, this time it will turn $60°$ counterclockwise, and vice versa.) Find the maximum length of ant's path.

4.8 St. Petersburg City Mathematical Olympiad (Russia)

Problem 9.1 Let $x_0 > x_1 > \cdots > x_n$ be real numbers. Prove that

$$x_0 + \frac{1}{x_0 - x_1} + \frac{1}{x_1 - x_2} + \cdots + \frac{1}{x_{n-1} - x_n} \geq x_n + 2n.$$

Problem 9.2 Let $f(x) = x^2 + ax + b$ be a quadratic trinomial with integral coefficients and $|b| \leq 800$. It is known also that $f(120)$ is prime. Prove that $f(x) = 0$ has no integer roots.

Problem 9.3 The vertices of a regular n-gon are labeled with $1, 2, \ldots, n$ with $n \geq 3$. For any three vertices A, B, C with $AB = AC$, the number at A is either larger than the numbers at B and C, or less than both of them. Find all possible values of n.

Problem 9.4 Points A_1, B_1, C_1 are chosen on the sides BC, CA, AB of an isosceles triangle ABC ($AB = BC$). It is known that $\angle BC_1A_1 = \angle CA_1B_1 = \angle A$. Let BB_1 and CC_1 meet at P. Prove that AB_1PC_1 is cyclic.

Problem 9.5 Find the set of possible values of the expression

$$f(x, y, z) = \left\{ \frac{xyz}{xy + yz + zx} \right\},$$

for positive integers x, y, z. Here $\{x\} = x - \lfloor x \rfloor$ is the fractional part of x.

Problem 9.6 Let AL be the angle bisector of triangle ABC. Parallel lines ℓ_1 and ℓ_2 equidistant from A are drawn through B and C respectively. Points M and N are chosen on ℓ_1 and ℓ_2 respectively such that AB and AC meet LM and LN at the midpoints of LM and LN respectively. Prove that $LM = LN$.

Problem 9.7 A *corner* is the figure resulted by removing 1 unit square from a 2×2 square. Prove that the number of ways to cut a 998×999 rectangle into corners, where two corners *can* form a 2×3 rectangle, does not exceed the number of ways to cut 1998×2000 rectangle into corners, so that *no* two form a 2×3 rectangle.

---

[1] Problems are numbered as they appeared in the contests. Repetition problems are skipped.

Problem 9.8 A convex $n-$gon $(n > 3)$ is divided into triangles by non-intersecting diagonals. Prove that one can mark $n-1$ segments among these diagonals and sides of the polygon so that no set of marked segments forms a closed polygon and no vertex belongs to exactly two segments.

Problem 10.1 The sequence $\{x_n\}$ of positive integers is formed by the following rule: $x_1 = 10^{999} + 1$, and for every $n \geq 2$, the number x_n is obtained from the number $11x_{n-1}$ by rubbing out its first digit. Is the sequence bounded?

Problem 10.2 Prove that any positive integer less than $n!$ can be represented as a sum of no more than n positive integer divisors of $n!$.

Problem 10.5 How many 10-digit numbers divisible by 66667 are there whose decimal representation contains only the digits $3, 4, 5$, and 6?

Problem 10.6 The numbers $1, 2, \ldots, 100$ are arranged in the squares of a 10×10 table in the following way: the numbers $1, \ldots, 10$ are in the bottom row in increasing order, numbers $11, \ldots, 20$ are in the next row in increasing order, and so on. One can choose any number and two of its neighbors in two opposite directions (horizontal, vertical, or diagonal). Then either the number is increased by 2 and its neighbors are decreased by 1, or the number is decreased by 2 and its neighbors are increased by 1. After several such operations the table again contains all the numbers $1, 2, \ldots, 100$. Prove that they are in the original order.

Problem 10.7 Quadrilateral $ABCD$ is inscribed in circle ω centered at O. The bisector of $\angle ABD$ meets AD and ω at points K and M respectively. The bisector of $\angle CBD$ meets CD and ω at points L and N respectively. Suppose that $KL \parallel MN$. Prove that the circumcircle of triangle MON goes through the midpoint of BD.

Problem 11.1 There are 150 red, 150 blue, and 150 green balls flying under the big top in the circus. There are exactly 13 green balls inside every blue one, and exactly 5 blue balls and 19 green balls inside every red one. Prove that some green ball is not contained in any of the other 449 balls.

Problem 11.3 Let $\{a_n\}$ be an arithmetic sequence of positive integers. For every n, let p_n be the largest prime divisor of a_n. Prove that the sequence $\{a_n/p_n\}$ is unbounded.

Problem 11.4 All positive integers not exceeding 100 are written on both sides of 50 cards (each number is written exactly once). The cards are put on a table so that only numbers on the top side may be seen. Vasya can choose several cards, turn them upside down, and then find the sum of all 50 numbers now on top. What is the maximum sum Vasya can be sure to obtain or beat?

Problem 11.5 Two players play the following game. They in turn write on a blackboard different divisors of 100! (except 1). A player loses if after his turn, the greatest common divisor of the all the numbers written becomes 1. Which of the players has a winning strategy?

Problem 11.7 A connected graph G has 500 vertices, each with degree 1, 2, or 3. We call a black-and-white coloring of these vertices *interesting* if more than half of the vertices are white but no two white vertices are connected. Prove that it is possible to choose several vertices of G so that in any interesting coloring, more than half of the chosen vertices are black.

Problem 11.8 Three conjurers show a trick. They give a spectator a pack of cards with numbers $1, 2, \ldots, 2n + 1$ ($n > 6$). The spectator takes one card and arbitrarily distributes the rest evenly between the first and the second conjurers. Without communicating with each other, these conjurers study their cards, each chooses an ordered pair of their cards, and gives these pairs to the third conjurer. The third conjurer studies these four cards and announces which card is taken by the spectator. Explain how such a trick can be done.

Glossary

AM–GM inequality If $a_1, a_2, \ldots, a_n$ are n nonnegative numbers, then

$$\frac{1}{n}\sum_{i=1}^{n} a_i \geq (a_1 a_2 \cdots a_n)^{\frac{1}{n}}$$

with equality if and only if $a_1 = a_2 = \cdots = a_n$.

Binomial coefficient

$$\binom{n}{k} = \frac{n!}{k!(n-k)!},$$

the coefficient of x^k in the expansion of $(x+1)^n$.

Binomial theorem

$$(x+y)^n = \sum_{k=0}^{n} \binom{n}{k} x^{n-k} y^k.$$

Cauchy–Schwarz inequality For any real numbers $a_1, a_2, \ldots, a_n$, and $b_1, b_2, \ldots, b_n$

$$\sum_{i=1}^{n} a_i^2 \cdot \sum_{i=1}^{n} b_i^2 \geq \left(\sum_{i=1}^{n} a_i b_i\right)^2,$$

with equality if and only if a_i and b_i are proportional, $i = 1, 2, \ldots, n$.

Centrally symmetric A geometric figure is centrally symmetric (centrosymmetric) about a point O if, whenever P is in the figure and O is the midpoint of a segment PQ, then Q is also in the figure.

Centroid of a triangle Point of intersection of the medians.

Centroid of a tetrahedron Point of the intersection of the segments connecting the midpoints of the opposite edges, which is the same as the point of intersection of the segments connecting each vertex with the centroid of the opposite face.

Ceva's theorem and its trigonometric form Let AD, BE, CF be three cevians of triangle ABC. Then AD, BE, CF are concurrent if and only if either

$$\frac{AF}{FB} \cdot \frac{BD}{DC} \cdot \frac{CE}{EA} = 1;$$

or

$$\frac{\sin \angle ABE}{\sin \angle EBC} \cdot \frac{\sin \angle BCF}{\sin \angle FCA} \cdot \frac{\sin \angle CAD}{\sin \angle DAB} = 1.$$

Cevian A cevian of a triangle is any segment joining a vertex to a point on the opposite side.

Circle of Apollonius Let A and B be two fixed points in the plane. The locus of the points P such that $PA : PB = k$ is constant is a circle, the circle of Apollonius.

Circumcenter Center of the circumscribed circle or sphere.

Circumcircle Circumscribed circle.

Congruence $a \equiv b \pmod{p}$, $a - b$ is divisible by p.

Concave up (down) function A function $f(x)$ is concave up (down) on $[a, b]$ if $f(x)$ lies under the line connecting $(a_1, f(a_1))$ and $(b_1, f(b_1))$ for all

$$a \leq a_1 < x < b_1 \leq b.$$

Cyclic polygon Polygon that can be inscribed in a circle.

Desargue's theorem If two triangles have corresponding vertices joined by lines which are either parallel or concurrent, then the intersections of corresponding sides are collinear. The converse holds: if the intersections of corresponding sides are collinear, then the lines joining corresponding vertices are either parallel or concurrent.

Euler's formula Let O and I be the circumcenter and incenter, respectively, of a triangle with circumradius R and inradius r. Then

$$OI^2 = R^2 - 2rR.$$

Euler line The orthocenter, centroid and circumcenter of any triangle are collinear. The centroid divides the distance from the orthocenter to the circumcenter in the ratio of $2:1$. The line on which these three points lie is called the Euler line of the triangle.

Euler's theorem If m is relatively prime to a, then $a^{\phi(m)} \equiv a \pmod{m}$, where $\phi(m)$ is the number of positive integers less than a and relatively prime to a.

Excircles or escribed circles Given a triangle ABC, there are four circles tangent to the lines AB, BC, CA. One is the inscribed circle, which lies in the interior of the triangle. One lies on the opposite side of line BC as A, and is called the excircle (escribed circle) opposite A, and similarly for the other two sides. The excenter opposite A is the center of the excircle opposite A; it lies on the internal angle bisector of A and the external angle bisectors of B and C.

Excenters See **excircles.**

Exradii The radii of the three excircles of a triangle.

Fermat's little theorem If p is prime, then $a^p \equiv a \pmod{p}$.

Feuerbach circle The feet of the three altitudes of any triangle, the midpoints of the three sides, and the midpoints of segments from the three vertices to the orthocenter, all line on the same circle, the Feuerbach cicle or the **nine-point circle** of the triangle. Let R be the circumradius of the triangle. The nine-point circle of the triangle has radius $R/2$ and is centered at the midpoint of the segment joining the orthocenter and the circumcenter of the triangle.

Feuerbach's theorem The nine point circle of a triangle is tangent to the incircle and the to the three excircles of the triangle.

Fibonacci sequence Sequence defined recursively by

$$F_1 = F_2 = 1, \qquad F_{n+1} = F_n + F_{n-1}.$$

Generating function If $a_0, a_1, a_2, \ldots$ is a sequence of numbers, then the generating function for the sequence is the infinite series

$$a_0 + a_1 x + a_2 x^2 + \cdots.$$

If f is a function such that

$$f(x) = a_0 + a_1 x + a_2 x^2 + \cdots,$$

then we also refer to f as the generating function for the sequence.

Harmonic conjugates Let A, C, B, D be four points on a line in that order. If the points C and D divide AB internally and externally in the same ratio, (i.e., $AC : CB = AD : DB$), then the points C and D are said to be the harmonic conjugates of each other with respect to the points A and B, and AB is said to be **harmonically divided** by the points C and D. If C and D are harmonic with respect to A and B, then A and B are harmonic with respective to C and D.

Harmonic range The four points A, B, C, D are referred to as a harmonic range, denoted by $(ABCD)$, if C and D are harmonic with respect to A and B.

Helly's theorem If $n > d$ and $C_1, \ldots, C_n$ are convex subsets of R^d, each $d+1$ of which have nonempty intersection, then there is a point in common to all the sets.

Hero's formula The area of a triangle with sides a, b, c is equal to

$$\sqrt{s(s-a)(s-b)(s-c)},$$

where $s = (a+b+c)/2$.

Homothety A homothety (central similarity) is a transformation that fixes one point O (its center) and maps each point P to a point P' for which O, P, P' are collinear and the ratio $OP : OP' = k$ is constant (k can be either positive or negative), where k is called the **magnitude** of the homothety.

Incenter Center of inscribed circle.

Incircle Inscribed circle.

Inversion of center O and ratio r Given a point O in the plane and a real number $r > 0$, the inversion through O with radius r maps every point $P \neq O$ to the point P' on the ray $\overrightarrow{OP}$ such that $OP \cdot OP' = r^2$. We also refer to this map as inversion through ω, the circle with center O and radius r. Key properties of inversion are:

1. Lines through O invert to themselves (though the invididual points on the line are not all fixed).
2. Lines not through O invert to circles through O and vice versa.
3. Circles not through O invert to other circles not through O.
4. A circle other than ω inverts to itself (as a whole, not point-by-point) if and only if it is orthogonal to ω, that is, it intersects ω and the tangents to the circle and to ω at either intersection point are perpendicular.

Jensen's inequality If f is concave up on an interval $[a, b]$ and λ_1, λ_2, ..., λ_n are nonnegative numbers with sum equal to 1, then

$$\lambda_1 f(x_1) = \lambda_2 f(x_2) + \cdots + \lambda_n f(x_n) \geq f(\lambda_1 x_1 + \lambda_2 x_2 + \cdots + \lambda_n x_n)$$

for any $x_1, x_2, \ldots, x_n$ in the interval $[a, b]$. If the function is concave down, the inequality is reversed.

Lattice point In the Cartesian plane, the lattice points are the points (x, y) for which x and y are both integers.

Law of sines In a triangle ABC with circumradius equal to R one has

$$\frac{\sin A}{BC} = \frac{\sin B}{AC} = \frac{\sin C}{AB} = 2R.$$

Lucas' theorem Let p be a prime; let a and b be two positive integers such that

$$a = a_k p^k + a_{k-1} p^{k-1} + \cdots a_1 p + a_0, b = b_k p^k + b_{k-1} p^{k-1} + \cdots b_1 p + b_0,$$

where $0 \leq a_i, b_i < p$ are integers for $i = 0, 1, \ldots, k$. Then

$$\binom{a}{b} \equiv \binom{a_k}{b_k}\binom{a_{k-1}}{b_{k-1}} \cdots \binom{a_1}{b_1}\binom{a_0}{b_0} \pmod{p}.$$

Matrix A matrix is a rectangular array of objects. A matrix A with m rows and n columns is an $m \times n$ matrix. The object in the i-th row and j-th column of matrix A is denoted $a_{i,j}$. If a matrix has the same number of rows as it has columns, then the matrix is called a square matrix. In a square $n \times n$ matrix A, the **main diagonal** consists of the elements $a_{1,1}, a_{2,2}, \ldots, a_{n,n}$.

Menelaus' theorem Let a transversal cut the sides BC, CA, AB of triangle ABC in F, G, H, respectively. Then

$$\frac{AH}{HB} \cdot \frac{BF}{FC} \cdot \frac{CG}{GA} = 1.$$

De Moivre's formula For any angle α and for any integer n,

$$(\cos\alpha + i\sin\alpha)^n = \cos n\alpha + i\sin n\alpha.$$

Nine point circle See **Feuerbach circle**.

Orthocenter of a triangle Point of intersection of altitudes.

Periodic function $f(x)$ is periodic with period $T > 0$ if

$$f(x+T) = f(x)$$

for all x.

Permutation Let S be a set. A permutation of S is a one-to-one function $\pi : S \to S$ that maps S onto S. If $S = \{x_1, x_2, \ldots, x_n\}$ is a finite set, then we may denote a permutation π of S by $\{y_1, y_2, \ldots, y_n\}$, where $y_k = \pi(x_k)$.

Pigeonhole principle If n objects are distributed among $k < n$ boxes, some box contains at least two objects.

Pole-Polar transformation Let C be a circle with center O and radius R. The pole-polar transformation with respect to C maps points different from O to lines, and lines that do not pass through O to points. If $P \neq O$ is a point then the **polar** of P is the line p' that is perpendicular to ray $\overrightarrow{OP}$ and satisfies

$$d(O,P)d(O,p') = R^2,$$

where $d(A,B)$ denote the distance between the objects A and B. If q is a line that does not pass through O, then the **pole** of q is the point Q' that has polar q. The pole-polar transformation with respect to the circle C is also called **reciprocation** in the circle C.

Polynomial in x of degree n Function of the form $f(x) = \sum_{k=0}^{n} a_k x^k$.

Power of a point Given a fixed point P and a fixed circle ω. Draw a line through P which intersects the circle at X and Y. The power of the point P with respect to ω is defined to be $PX \cdot PT$. The **power of a point theorem** states that this quantity is a constant; i.e., does not depend on which line was drawn. Note that it did not matter if P was in, on or outside ω.

Ptolemy's theorem In a cyclic quadrilateral $ABCD$,

$$AC \cdot BD = AB \cdot CD + AD \cdot BC.$$

Pythagorean triple A set (a, b, c) of three numbers is a Pythagorean triple if there is a right-angled triangle with sides of lengths a, b, c. If c is the length of the hypotenuse, this is equivalent to the assertion:

$$c^2 = a^2 + b^2.$$

If a, b, c are integers, the triple is **primitive** if the greatest common divisor of a, b, c is 1. All primitive Pythagorean triples (a, b, c) are given parametrically by

$$a = 2uv, \qquad b = u^2 - v^2, \qquad c = u^2 + v^2,$$

where $u > v$ are relatively prime integers.

Radical axis Let ω_1 and ω_2 be two non-concentric circles. The locus of all points of equal power with respect to these circles is called the radical axis of ω_1 and ω_2. Let $\omega_1, \omega_2, \omega_3$ be three circles whose centers are not collinear. There is exactly one point whose powers with respect to the three circles are all equal. This point is called the **radical center** of $\omega_1, \omega_2, \omega_3$.

Root of an equation Solution to the equation.

Root of unity Solution to the equation $z^n - 1 = 0$.

Root Mean Square-Arithmetic Mean Inequality For positive numbers $x_1, x_2, \ldots, x_n$,

$$\sqrt{\frac{x_1^2 + x_2^2 + \cdots + x_k^2}{n}} \geq \frac{x_1 + x_2 + \cdots + x_k}{n}.$$

More generally, let $a_1, a_2, \ldots, a_n$ be any positive numbers for which $a_1 + a_2 + \cdots + a_n = 1$. For positive numbers $x_1, x_2, \ldots, x_n$ we define

$$M_{-\infty} = \min\{x_1, x_2, \ldots, x_k\},$$
$$M_{\infty} = \max\{x_1, x_2, \ldots, x_k\},$$
$$M_0 = x_1^{a_1} x_2^{a_2} \cdots x_n^{a_n},$$
$$M_t = (a_1 x_1^t + a_2 x_2^t + \cdots + a_k x_k^t)^{1/t},$$

where t is a non-zero real number. Then

$$M_{-\infty} \leq M_s \leq M_t \leq M_{\infty}$$

for $s \leq t$.

Triangular number A number of the form $n(n+1)/2$, where n is some positive integer.

Trigonometric identities

$$\sin^2 x + \cos^2 x = 1,$$

$$\tan x = \frac{\sin x}{\cos x},$$

$$\cot x = \frac{1}{\tan x}$$

addition and subtraction formulas:

$$\sin(a \pm b) = \sin a \cos b \pm \cos a \sin b,$$

$$\cos(a \pm b) = \cos a \cos b \mp \sin a \sin b,$$

$$\tan(a \pm b) = \frac{\tan a \pm \tan b}{1 \mp \tan a \tan b};$$

double-angle formulas:

$$\sin 2a = 2 \sin a \cos a,$$

$$\cos 2a = 2\cos^2 a - 1 = 1 - 2\sin^2 a,$$

$$\tan 2a = \frac{2\tan a}{1 - \tan^2 a},$$

triple-angle formulas:

$$\sin 3a = 3\sin a - 4\sin^3 a,$$

$$\cos 3a = 4\cos^3 a - 3\cos a,$$

$$\tan 3a = \frac{3\tan a - \tan^3 a}{1 - 3\tan^2 a};$$

half-angle formulas:

$$\sin a = \frac{2\tan\frac{a}{2}}{1 + \tan\frac{a}{2}},$$

$$\cos a = \frac{1 - \tan^2\frac{a}{2}}{1 + \tan\frac{a}{2}},$$

$$\tan a = \frac{2\tan\frac{a}{2}}{1 - \tan^2\frac{a}{2}};$$

sum-to-product formulas:

$$\sin a + \sin b = 2\sin\frac{a+b}{2}\cos\frac{a-b}{2},$$

$$\cos a + \cos b = 2\cos\frac{a+b}{2}\cos\frac{a-b}{2},$$

$$\tan a + \tan b = \frac{\sin(a+b)}{\cos a\cos b};$$

difference-to-product formulas:

$$\sin a - \sin b = 2\sin\frac{a-b}{2}\cos\frac{a+b}{2},$$

$$\cos a - \cos b = -2\sin\frac{a-b}{2}\sin\frac{a+b}{2},$$

$$\tan a - \tan b = \frac{\sin(a-b)}{\cos a\cos b};$$

product-to-sum formulas:

$$2\sin a\cos b = \sin(a+b) + \sin(a-b),$$

$$2\cos a\cos b = \cos(a+b) + \cos(a-b),$$

$$2\sin a\sin b = -\cos(a+b) + \cos(a-b).$$

Classification of Problems

Algebra

Belarus	99-R4-10.1[1], 11.1; 99-S-1[2], 6
Bulgaria	98-3, 4, 8, 10, 18
Canada	98-2[3]
China	99-2
Czech and Slovak Republics	98-5; 99-6
Hungary	98-3; 99-5, 11, 14, 17
India	98-2, 9, 12, 13
Iran	98-11; 99-R2-1, 3
Ireland	98-9; 99-5, 6
Italy	99-9
Japan	99-4
Korea	98-1; 99-2, 4
Poland	98-2; 99-4
Romania	98-9, 11, 12; 99-8.1, 9.2, 9.4, 10.1, 10.3; 99-S-9
Russia	98-9, 20, 22, 29, 37, 42; 99-R4-8.1, 11.1, 11.8; 99-R5-10.2
Turkey	98-2; 99-7
U.K.	98-9; 99-3, 5
Ukraine	99-1
Vietnam	98-1, 6; 99-1, 3, 6, 10

[1] 1999 round 4 problem 10.1
[2] 1999 IMO Selection Test problem 1
[3] 1998 problem 2

Combinatorics

Balkan	98-3
Czech-Slovak Match	98-4
Hong Kong	99-2, 3
Iberoamerican	98-3, 4; 99-3
Olimpiada de mayo	99-3
St. Petersburg	98-3, 5, 6, 7, 8, 9, 11, 13, 17, 18, 19, 21, 27, 28; 99-9.3, 9.7, 9.8, 10.6, 11.1, 11.4, 11.5, 11.6

Geometry

Belarus	99-R4-10.4, 10.5, 11.4, 11.7; 99-S-5, 7, 8
Brazil	99-1
Bulgaria	98-2, 5, 9, 14 99-R3-5; 99-R4-5
Canada	98-4; 99-2
China	98-1; 99-1
Czech and Slovak Republics	98-3, 4; 99-2, 3, 5
France	99-5
Hungary	98-2, 4, 7, 10; 99-2, 3, 9, 10
India	98-4, 6
Iran	98-1, 4, 7, 16; 99-R1-3, 4, 5; 99-R2-4; 99-R3-2
Ireland	98-2; 99-3, 10
Italy	99-1
Korea	98-5; 99-1
Poland	98-3, 5; 99-6
Romania	98-5; 99-7.1, 7.4, 8.3, 8.4, 9.1, 9.3, 99-S-2, 10
Russia	98-2, 5, 6, 10, 14, 18, 34, 39, 47; 99-R4-8.3, 8.6, 9.2, 9.8, 10.2, 10.6; 99-R5-9.3, 9.7, 10.3, 11.3, 11.6, 11.7
Turkey	98-1, 5; 99-1, 4, 6
U.K.	98-3, 5, 7; 99-2, 7
U.S.A.	98-2; 99-6
Ukraine	99-3
Vietnam	99-2, 5, 8, 11
Asian Pacific	98-4; 99-3
Austrian-Polish	98-6, 9; 99-4, 8

Geometric Inequalities

Inequalities

Iran	98-3, 9, 14; 99-R1-1; 99-R3-4, 5
Ireland	98-1, 7; 99-1, 8
Korea	98-4; 99-6
Poland	99-5
Romania	99-8.2, 10.4; 99-S-4, 5, 7
Russia	98-28; 99-R4-9.3, 11.5; 99-R5-9.6, 10.7
Taiwan	98-5
Turkey	99-2
U.K.	99-8
U.S.A.	98-3; 99-4
Ukraine	99-2
Vietnam	98-4, 5; 99-4, 7, 9
Asian Pacific	98-3; 99-2
Austrian-Polish	98-1, 4; 99-2
Balkan	98-2; 99-4
Czech-Slovak Match	99-1, 6
St. Petersburg	99-9.1

Number Theory

Belarus	99-R4-10.2, 10.6, 11.2, 11.3, 11.6; 99-S-2, 3, 11.
Bulgaria	98-6, 11, 13 99-R3-1, 4; 99-R4-4, 6
Canada	98-1, 5; 99-1, 3
China	98-4; 99-4
Czech and Slovak Republics	98-1; 99-1
France	99-2
Hungary	98-5, 6; 99-1, 4, 8, 16
India	98-1, 5, 7
Iran	98-5, 6; 99-R1-5
Ireland	98-3, 5, 6; 99-2, 7, 9
Italy	99-2, 5, 6, 7
Japan	98-1; 99-2
Korea	98-1, 3; 99-3
Poland	98-1, 4; 99-3
Romania	98-7; 99-7.2; 99-S-1, 3, 6
Russia	98-1, 7, 26, 31; 99-R4-8.2, 8.5, 9.1, 9.7, 10.5, 10.8; 99-R5-9.1, 10.5, 11.1, 11.5

Miscellaneous